T0269581

CAMBRIDGE LIBRARY COLLECTION

Books of enduring scholarly value

Technology

The focus of this series is engineering, broadly construed. It covers technological innovation from a range of periods and cultures, but centres on the technological achievements of the industrial era in the West, particularly in the nineteenth century, as understood by their contemporaries. Infrastructure is one major focus, covering the building of railways and canals, bridges and tunnels, land drainage, the laying of submarine cables, and the construction of docks and lighthouses. Other key topics include developments in industrial and manufacturing fields such as mining technology, the production of iron and steel, the use of steam power, and chemical processes such as photography and textile dyes.

An Historical and Descriptive Account of the Steam Engine

Though much about his life is uncertain, Charles Frederick Partington is known to have lectured at the London Institution between 1823 and 1830 on a variety of technical topics, and he delivered some of the first lectures specifically designed for young people. He had a particular interest in the steam engine, and this book, reissued here in the first edition of 1822, was one of the earliest overviews of its history and development. A third edition appeared in 1826. Noting the excessive frequency with which 'the faults of any new invention are unjustly magnified, while its real advantages are seldom duly appreciated', the author is keen to act as evangelist. Detailed and illustrated descriptions of various early engines are included, comparing their characteristics and advantages. Also of note are Partington's descriptions of early attempts to mitigate the 'smoke and noxious effluvia which proceed from their capacious vomitories'.

Cambridge University Press has long been a pioneer in the reissuing of out-of-print titles from its own backlist, producing digital reprints of books that are still sought after by scholars and students but could not be reprinted economically using traditional technology. The Cambridge Library Collection extends this activity to a wider range of books which are still of importance to researchers and professionals, either for the source material they contain, or as landmarks in the history of their academic discipline.

Drawing from the world-renowned collections in the Cambridge University Library and other partner libraries, and guided by the advice of experts in each subject area, Cambridge University Press is using state-of-the-art scanning machines in its own Printing House to capture the content of each book selected for inclusion. The files are processed to give a consistently clear, crisp image, and the books finished to the high quality standard for which the Press is recognised around the world. The latest print-on-demand technology ensures that the books will remain available indefinitely, and that orders for single or multiple copies can quickly be supplied.

The Cambridge Library Collection brings back to life books of enduring scholarly value (including out-of-copyright works originally issued by other publishers) across a wide range of disciplines in the humanities and social sciences and in science and technology.

An Historical and Descriptive Account of the Steam Engine

Comprising a General View of the Various Modes of Employing Elastic Vapour as a Prime Mover in Mechanics

Charles Frederick Partington

CAMBRIDGE
UNIVERSITY PRESS

University Printing House, Cambridge, CB2 8BS, United Kingdom

Cambridge University Press is part of the University of Cambridge.
It furthers the University's mission by disseminating knowledge in the pursuit of
education, learning and research at the highest international levels of excellence.

www.cambridge.org
Information on this title: www.cambridge.org/9781108071031

This edition first published 1822
This digitally printed version 2014

ISBN 978-1-108-07103-1 Paperback

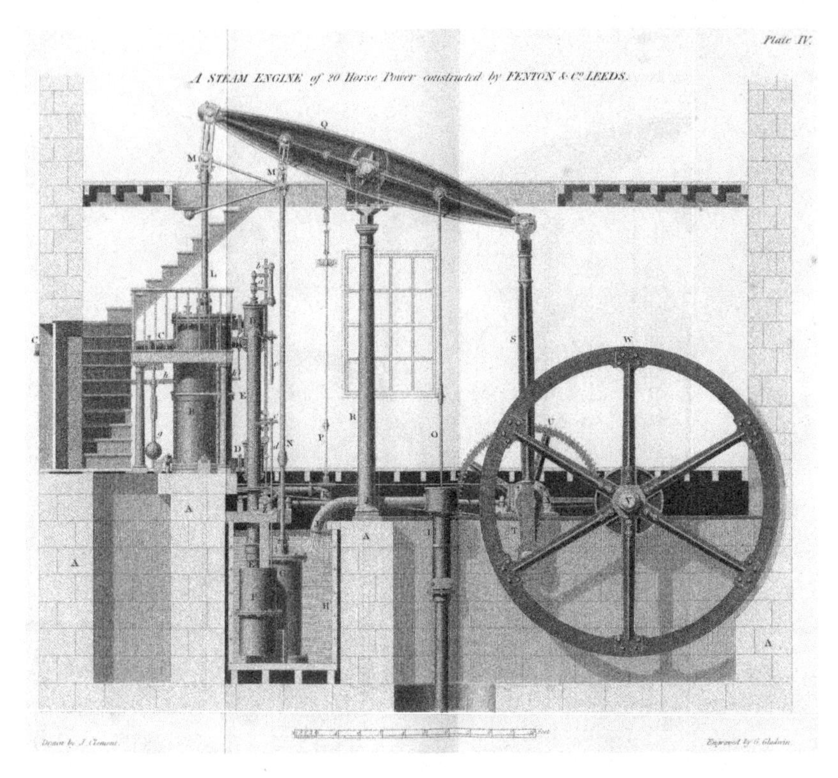

The material originally positioned here is too large for reproduction in this reissue. A PDF can be downloaded from the web address given on page iv of this book, by clicking on 'Resources Available'.

AN HISTORICAL AND DESCRIPTIVE ACCOUNT OF THE STEAM ENGINE,

COMPRISING A GENERAL VIEW OF THE VARIOUS MODES OF EMPLOYING ELASTIC VAPOUR AS A PRIME MOVER IN MECHANICS;

WITH

AN APPENDIX

OF PATENTS AND PARLIAMENTARY PAPERS CONNECTED WITH THE SUBJECT.

BY CHARLES FREDERICK PARTINGTON,
OF THE LONDON INSTITUTION.

"Soon shall thy arm, unconquer'd Steam! afar!
Drag the slow barge, or drive the rapid car;
Or on wide-waving wings expanded bear
The flying chariot through the fields of air." *Darwin.*

Illustrated by Thirteen Engravings and Diagrams.

London:
PRINTED FOR J. TAYLOR,
AT THE ARCHITECTURAL LIBRARY, 59, HIGH HOLBORN.

1822.

J. and T. Bartlett, Printers, Oxford.

TO

WILLIAM HASLEDINE PEPYS, Esq.

FELLOW OF THE ROYAL AND LINNEAN SOCIETIES; A VICE-PRESIDENT OF THE GEOLOGICAL SOCIETY; MEMBER OF THE ROYAL INSTITUTION, OF THE ASTRONOMICAL SOCIETY, &c. &c. AND HON. SECRETARY OF THE LONDON INSTITUTION:

As a Tribute of Respect for his high Scientific Attainments, and liberal encouragement of Philosophical Learning; this Historical Account of one of the most valuable applications of Science to the Arts and Manufactures of a great Commercial Nation, is, with his permission, respectfully inscribed,

By his much obliged,

and faithful humble Servant,

CHARLES FREDERICK PARTINGTON.

INTRODUCTION.

THE great practical use of machinery to a commercial country is so well known, and its superiority to animal force so universally acknowledged and felt in every branch of our manufactures, that but little apology will be necessary for introducing to the man of science and practical artizan, a work, the avowed object of which is, to render the uses and general principles of the Steam Engine familiar to every class of persons. That it has enabled England to support a proud pre-eminence, both in arts and political power, is equally apparent; and it is a fact much to be deplored, that while some of the least important of the arts connected with domestic life, have been illustrated and explained by men celebrated for scientific research, a description and account of the uses of this stupendous machine, have been left to the Cyclopædias and other works of a general nature.

The principal object to be attained by the employment of the Steam Engine, as well as every other species of machinery, being the reduction of animal labour, it may be advisable, before we proceed to the more immediate subject of the present work, to compare the various species of artificial power that have hitherto been employed for that purpose; and by this method we may be enabled to calculate with certainty and precision, on the most economical mode of producing a given force. To form, however, an accurate estimate of the saving thus effected, it will be necessary to examine, though but briefly, the amount of animal force and its result as applied to machinery.

From the most accurate observations, it appears, that the physical powers of the human race differ very widely, not only in various individuals, but also in different climates; the value of a man therefore, as a working machine, will not be so great beneath the torrid zone as in the more temperate climate of Europe. This will serve to illustrate the great advantage which our Colonists, particularly in the West Indies, would derive from the more general employment of inanimate force; the day labour of a negro in the sugar countries, amounting to little more than one-third of that performed by an European mechanic.

A labourer working ten hours per day, can raise in one minute a weight equivalent to 3750 pounds one foot high, or about sixty cubic feet of water in the same time; while the power of a horse working eight hours per day, may be correctly averaged at 20,000 pounds. Smeaton states, that this animal, by means of pumps, can raise two hundred and fifty hogsheads of water ten feet high in an hour. It is a well known fact, also, that men when trained to running, are able on the average of several days being taken, to outstrip the fleetest horse; and yet it will be seen from the above statement, that his force, if properly applied, is at least six times that of the most powerful man.

The use of water as an impelling power, both for the turning of machinery and other purposes connected with the useful arts, appears to have been known at a very early period. Vitruvius describes a variety of machines for this purpose, the earliest of which were employed merely to raise a portion of the fluid by which they were impelled. The most simple method of applying this element as a mechanical agent, evidently consisted in the construction of a wheel, the periphery of which was composed of a number of float-boards. This, on being exposed to the action of a running stream, was afterwards employed to give motion to a variety of mills, and is at the present time employed in almost every species of machinery.

Among the most celebrated hydraulic machines, we may enumerate the Machine of Marly. This, when first constructed, appears to have produced one-eighth of the power expended, so that seven-eighths of its power were usually lost. This misapplied power has been injurious to the engine; and the wear it has occasioned, has reduced the mechanical effect very materially. But this may be considered as an extreme case, and we select it merely as an instance of that total ignorance of the first principles of mechanics, which characterized some foreign engineers of the last century.

It may, however, be advisable to examine the ratio of power expended in comparison with that of the effect produced in some of the most simple hydraulic machines; and by this calculation, the amount of friction, &c. may be accurately ascertained.

	Power		Effect
Undershot water wheel	9	=	3
Overshot do.	10	=	8
Hydraulic Ram. (This machine will make from 20 to 100 strokes per minute.)	10	=	6
Large machine at Chremnitz, (each stroke occupying about three minutes.)	9	=	3

But the water-mill, which is the usual machine employed, even in its most improved form, is far from being beneficial either to the agriculturist or the manufacturer. The former is injured by the laws which prohibit the draining of mill-streams for the purposes of irrigation, by which much improvement is kept back that would otherwise take place; while the health of the latter, in the immediate neighbourhood of manufacturing districts, is much injured by the stagnant condition of the water which is thus unnecessarily dammed up.

Wind, which we may consider as the next substitute for animal power, appears to have been first employed to give motion to machinery in the beginning of the sixth century.

The use of this species of mechanic force, is however principally limited to the grinding of corn, the pressing of seed, and other simple manipulations; the great irregularity of this element precluding its application to those processes which require a continued motion.

A windmill with four sails, measuring seventy feet from the extremity of one sail, to that of the opposite one, each being six feet and an half in width, is capable of raising 926 pounds, two hundred and thirty-two feet in a minute; and of working on an average eight hours per day. This is equivalent to the work of thirty-four men; twenty-five square feet of canvas performing the average work of a day labourer. A mill of this magnitude seldom requires the attention of more than two men; and it will thus be seen, that making allowance for its irregularity, wind possesses a decided superiority over every species of animal labour.

To shew, however, the great advantage the Steam Engine, even in its rudest state, possesses over mere pneumatic or hydraulic machinery, we will now examine its effective force when employed in the working of pumps. It has been already stated, that the Machine of Marly, formerly considered the most powerful engine in the world, when first erected lost seven-eighths of its power from friction, and other causes; while the overshot water-wheel which can act only in favourable situations, produces nearly eight-tenths of the force employed. Now it is stated by Dr. Desaguliers, that the atmospheric engine working at Griff-mine, nearly a century back, produced full two-thirds of effective force for the power employed; and this too at a comparatively moderate expense. We find, farther, that an hundred weight of coals burned in an engine on the old construction, would raise at least twenty thousand cubic feet of water twenty-four feet high; an engine with a twenty-four-inch cylinder doing the work of seventy-four horses. From this it will be seen that a bushel of coals is equal to two horses, and that every inch of the cylinder performs nearly the work of a man.

An engine upon Captain Savery's plan, constructed by Mr. Keir, has been found to raise nearly three millions of pounds of water one foot high with a single bushel of coals; while the best engine on Newcomen's principle will raise ten millions, and Mr. Watt's engine upwards of thirty millions of pounds, the same height. If we add to the advantage gained by the employment of so cheap a prime mover, the vast concentration of force thus brought into immediate action, its value may easily be appreciated.

One of the largest engines yet constructed, is now in action at the United Mine in Cornwall, it raises eighty thousand pounds one hundred feet in height per minute: and to effect this enormous labour, it only requires about thirty pounds of coal for the same period of time.

To the mining interests this valuable present of Science to the Arts has been peculiarly acceptable; as a large portion of our now most productive mineral districts must have long ere this been abandoned, had not the Steam Engine been employed as an active auxiliary in those stupendous works. In draining of fens and marsh lands this machine is in the highest degree valuable; and in England, particularly, it might be rendered still more generally useful. In practice it has been ascertained that an engine of six-horse power, will drain more than eight thousand acres, raising the water six feet in height;

while the cost of erection for an engine for this species of work, including the pumps, will not exceed seven hundred pounds. This is more than ten windmills can perform, at an annual expenditure of several hundred pounds; while, in the former case, the outgoings will not exceed one hundred and fifty pounds per annum.

To the mariner, also, the Steam Engine offers advantages of a no less important and novel nature than those we have already described. By its use he is enabled to traverse the waters, both against wind and tide, with nearly as much certainty, and, as the machinery is now constructed, with much less danger, than by the most eligible road conveyance.

In proof of the speed of these vessels, it may be sufficient to state that the passage from or to London and Margate, which is more than eighty miles, by water, is often performed in the short space of six or seven hours! It too frequently, however, happens that the faults of any new invention are unjustly magnified, while its real advantages are seldom duly appreciated; and this axiom has been fully verified, in the clamour so unjustly raised against the application of the Steam

Engine to nautical purposes. Accidents are now, however, of but rare occurrence; and it is more than probable, that the great improvements that have been made in the boiler and safety-valve, will effectually secure these parts of the engine from a recurrence of those tremendous explosions, which unfortunately characterized the first introduction of Steam Navigation.

And, lastly, the political economist must hail with the most heartfelt gratification, the introduction of so able and efficient a substitute for animal labour as the Steam Engine. It has been calculated that there are at least ten thousand of these machines at this time at work in Great Britain; performing a labour more than equal to that of two hundred thousand horses, which, if fed in the ordinary way, would require above one million acres of land for subsistence; and this is capable of supplying the necessaries of life to more than fifteen hundred thousand human beings.

CONTENTS.

HISTORICAL ACCOUNT

OF THE

STEAM ENGINE.

CHAP. I.

Nature of Steam—Application of it as a moving power—Brancas—Marquis of Worcester—Sir Samuel Morland—Papin—Savery—Newcomen—Hulls—Falck—Amontons—Deslandes—Francois.

As the whole power of the Steam Engine depends on the employment of elastic vapour, produced from water at different temperatures, varying from 212°, or the boiling point of Fahrenheit's thermometer, to 300° of the same scale, it may be advisable in the first instance to examine some of the principal phenomena connected with the formation of vapour in its most simple form, and its application to the steam engine will then be sufficiently obvious.

Steam is highly rarefied water, the particles of which are expanded by the absorption of caloric, or the matter of heat. Water rises in vapour at all temperatures, though this is usually supposed to take place only at the boiling point: when, however, the evaporation occurs below 212°, it is confined to the surface of the fluid acted upon, but at that heat, steam is formed at the bottom of the water, and ascends through it, preventing its elevation to a higher temperature, by carrying off the heat in a latent form. At the common pressure of the atmosphere, one cubic inch of water produces about 1700 cubic inches of aqueous vapour or steam; but the boiling point, as we have already stated, varies very considerably, and these anomalies materially effect the density of the vapour produced. Thus, in a vacuum water boils at about 70°, under common pressure, at 212°; and when pressed by a column of mercury, five inches in height, water does not boil until it is heated to 217°; each inch of mercury producing by its pressure, a rise of about 1° in the thermometer.

According to Dr. Ure's elaborate experiments, the elastic force of this vapour at 212° is such, that it is equivalent to the pressure of a column of mercury 30 inches in height; at 226.3°, to that of 40 inches; at 238.5°, to 50.3 inches; at 257.5°, to 69.8 inches; at 273.7°, to 91.2 inches; at 285.2°, to 112.2 inches; at 312°, to 166 inches; and Mr. Woolf has ascertained that at these temperatures, omitting the last, a cubic foot of steam will expand to about five, ten, twenty, thirty, and forty times its volume respectively; its elastic force, when thus dilated, being in each case equal to the ordinary pressure of the atmosphere.

One pound of Newcastle coal converts seven pounds of boiling water into steam; and the time required to convert a given quantity of boiling water into steam, is six times that required to raise it from the freezing to the boiling point.

To shew by direct experiment the great expansive force of vapour from water when converted into steam, by the application of heat, it will only be necessary to take a glass tube, at one end of which is a bulb of two inches in diameter, and dropping into it a single spherule of water, the diameter of which will scarcely exceed one tenth of an inch, or about the eighteen hundredth part of the size of the glass bulb, we shall find, that it may very readily be expanded by the application of heat, so as to expel the air from the vessel. That this is actually the case, may be shewn by merely plunging the mouth of the tube into cold water, and suffering the steam to return to its original state, which being effected by the abstraction of a portion of its artificial heat, the water will rush in from the external vessel, and occupy the place of the steam thus condensed, which could not have taken place had any portion of the air remained in the tube or its bulb.

From these data, it will be evident, that when steam is merely employed to displace the air in a close vessel, and afterwards produce a vacuum by condensation, no more heat is necessary than what will raise the water for this purpose to 212°; but if, on the contrary, high pressure steam is required, a very considerable increase of heat will be essential; and of this kind was the elastic vapour employed in all the early steam engines to which we may now more immediately direct the reader's attention.

Among the numerous competitors for the honour of having first suggested steam as a moving power in mechanics, we must certainly place Brancas and the Marquis of Worcester in the foremost rank. The former of these was an Italian philosopher, of considerable eminence, and who, in 1629, published a treatise entitled, "Le Machine, &c." which contained a description of a machine for this purpose. The apparatus employed by Brancas, was in fact nothing more than a large æolipile, similar to the blow-pipe invented by M. Pictet of Geneva, with this difference, that the aperture in the pipe connected with the body of the æolipile instead of being directed to the lamp, (or in this case, the furnace that heated the machine,) was made to strike against the floats or vanes of a wheel, by which means a rotatory motion was produced.

After the publication of this scheme, which it is probable was never put in practice with any useful effect, nearly thirty years elapsed ere the farther consideration of this important subject was resumed by the Marquis of Worcester. The mode of employing steam recommended by the Marquis, and which he describes in his "Century of Inventions" to have completely carried into effect, was entirely different from that of his predecessor; and it is evident that the noble author had received no previous hint of Brancas' invention, as he expressly states in another part of the above work, that he "desired not to set down any other mens' inventions;" and if he had in any case acted on them, "to nominate likewise the inventor.*"

It is said that the Marquis, while confined in the

---

* This work was written about the middle of the seventeenth century, and considered as a description of the united discoveries of one individual, is certainly one of the most extraordinary scientific productions which has yet issued from the press in any age or nation. In addition, however, to its value, as containing the first tangible suggestion for the employment of steam, as an hydraulic and pneumatic force, it has unquestionably formed the foundation of a large portion of the patent inventions, which make so prominent a feature in the present day. The praiseworthy labours, however, of this indefatigable nobleman, shared the fate which usually attends on projections; and it was left to the slow though certain march of scientific improvement, to award to his memory a posthumous praise. The Marquis also published a work, entitled, "An exact and true Definition of the most stupendous Water-commanding Engine, invented by the Right Honourable (and deservedly to be praised and admired) Edward Somerset, Lord Marquis of Worcester, and by his Lordship himself presented to his most excellent Majesty Charles the Second, our most gracious Sovereign." This was published in a small quarto volume of only twenty-two pages, and consists of little more than an enumeration of the wonderful

Tower of London, was preparing some food on the fire of his apartment, and the cover having been closely fitted, was, by the expansion of the steam, suddenly forced off and driven up the chimney. This circumstance attracting his attention, led him to a train of thought, which terminated in this important discovery. But no figure has been preserved of his invention; nor, as we have good reason to suppose, any description of the machine he employed, except the sixty-eighth article in the above-mentioned work. We shall content ourselves, therefore, with extracting that article from the noble author's MS. preserved in the British Museum.

"An admirable and most forcible way to drive up water by fire; not by drawing or sucking it upwards, for that must be as the philosophers call it, *infra sphæram activitatis*, which is but at such a distance. But this way hath no boundary, if the vessels be strong enough; for I have taken a piece of a whole cannon, whereof the end was burst, and filled it three quarters full of water, stopping and screwing up the broken end, as also the touch-hole; and making a constant fire under it, within twenty-four hours it burst, and made a great crack; so that having found a way to make my vessels, so that they are strengthened by the

---

properties of the above engine; and it is certain that he never published any key to the first hint furnished in the *Century of Inventions*.

force within them, and the one to fill after the other, I have seen the water run like a constant fountain stream, forty feet high; one vessel of water, rarified by fire, driveth up forty of cold water. And a man that tends the work is but to turn two cocks, that one vessel of water being consumed, another begins to force and refill with cold water, and so successively, the fire being tended and kept constant, which the self-same person may likewise abundantly perform in the interim, between the necessity of turning the said cocks." Vide Harleian MSS. No. 2428.

In 1683, a scheme for raising water by the agency of steam was offered to the notice of Louis XIV. by an ingenious English mechanic, of the name of Morland; this, however, was evidently formed upon the plan previously furnished by the Mārquis of Worcester, in his Century of Inventions. Morland was presented to the French monarch in 1682, and in the course of the following year his apparatus is said to have been actually exhibited at St. Germain's.* The only notice of this plan

---

* Sir Samuel Morland was the son of a baronet of the same name, created by King Charles II. for his zealous services performed during the King's exile. The son was made *Magister Mechanicorum* by the King in 1681, and was justly celebrated at that period for a number of very ingenious inventions, among which we may enumerate the drum capstan for weighing anchors, the speaking trumpet, and fire engine. The celebrated John Evelyn gives the following account of a visit paid him at a very late period of his life:—

" The Abp. and myselfe went to Hammersmith, to visite Sir

occurs in the collection of MSS. to which we have already alluded, and forms the latter part of a very beautiful volume, containing about thirty-eight pages, and entitled " *Elevation des Eaux, par toute sorte de Machines, réduite a la mésure, au poids, et a la balance. Presentée a sa Majesté tres Chrestienne, par le Chevalier Morland, gentilhomme ordinaire de la chambre privée, et maistre des méchaniques du Roy de la Grande Brétaigne*, 1683."

The MS. is written upon vellum, richly illuminated, and the part which has reference to the steam engine occupies only four pages, commencing with a separate title, &c. It is also accompanied by a table of the sizes of cylinders, and the amount of water to be raised by a given force of steam. This curious memoir forms an important link in the chain of historical evidence,

---

Sam. Morland, who was entirely blind, a very mortifying sight. He shewed us his invention of writing, which was very ingenious, also his wooden kalender, which instructed him all by feeling, and other pretty and useful inventions of mills, pumps, &c. and the pump he had erected that serves water to his garden and to passengers, with an inscription, and brings from a filthy part of the Thames neere it a most perfect and pure water. He had newly buried 200*l*. worth of music books six feet under ground, being, as he said, love songs and vanity. He plays himselfe psalms and religious hymns on the *Theorbo*." Diary, Oct. 25th, 1695.

About the year 1684, Sir Samuel purchased a house at Hammersmith, and it appears from the register of that parish, he was buried Jan. 6th, 1696.

which tends to prove that the English, though not the actual inventors of the steam engine, were unquestionably the first to apply its stupendous powers to any useful practical purpose; we shall, therefore, offer no apology for presenting it entire to the notice of the reader.

" *Les Principes de la nouvelle Force de Feu: inventée par le Chevalier Morland, l'an.* 1682, *et presentée a sa Majesté tres Chrestienne*, 1683,

" L'eau etant evaporée par la force de feu, ces vapeurs demandent incontinent une plus grand'espace [environ deux mille fois] que l'eau n'occupoit auparavant, et plus lost que d'etre toujours emprisonnées, feroient crever une piece de canon. Mais etant bien gouvernées selon les regles de la statique, et par science réduites a la mésure, au poids et a la balance, alors elles portent paisiblement leurs fardeaux, [comme des bons chevaux,] et ainsi servoient elles du grand usage au gendre humain, particulierement pour l'elevation des eaux, selon la table suivante, qui marque les nombres des livres qui pourront etre levées 1800 fois par heure, a six pouces de louée, par de cylindres a moitié remplies d'eau, aussi bien que les divers diametres et profondeurs des dits cylindres."

Table of the Diameter and Length of Steam Cylinders; with the Number of Pounds Weight to be raised.

CYLINDERS.		
Diam. en Pieds.	Prof. en Pieds.	Livres du Poids, pour être leves.
1	2	15
2	4	120
3	6	405
4	8	960
5	10	1875
6	12	3240
Nombres des Cylindres, qui ont pour Diametre 6 Pieds, et 12 Pieds de profondeur.		
	1	3240
	2	6480
	3	9720
	4	12960
	5	16200
	6	19440
	7	22680
	8	25920
	9	29160
	10	32400
	20	64800
	30	97200
	40	129600
	50	162000
	60	194400
	70	226800
	80	259200
	90	291600

The invention of the atmospheric engine, though usually ascribed to Newcomen, or his coadjutor Savery, is unquestionably of French origin. An account of it having been published twelve years prior to the commencement of Newcomen's patent.

In 1695, Papin, then resident at Cassel, published a work, describing a variety of methods for raising water in which he enumerates the above invention. Being unable to procure this tract, we insert the following translation of that part which relates to the steam engine. It occurs in the Transactions of the Royal Society, for 1697. After alluding to the inconvenience of forming a vacuum by means of gunpowder, which was one of his early propositions, he recommends "the alternately turning a small surface of water into vapour, by fire applied to the bottom of the cylinder that contains it, which vapour forces up the plug in the cylinder to a considerable height, and which (as the vapour condenses, as the water cools when taken from the fire) descends again by the air's pressure, and is applied to raise the water out of the mine."

From this it will be evident that any practical mechanic would have suggested the further application of pumps and a working beam or lever similar to those in Newcomen's engine.

To experimentally illustrate the principle on which the steam or atmospheric engine acts, we have only to procure an hollow bulb of glass, connected with a tube of the same material, about four or five inches in length, and furnished with a piston or plug, sliding air-tight. A small quantity of water being placed in the vessel, must then be heated to the boiling point, and the vapour formed will speedily impel the piston to the open end. The bulb must now be withdrawn from the candle, and on being immersed in a vessel of cold water the vapour will rapidly condense; and the minute particles of which it is composed will return to their original bulk. A vacuum being thus formed within the vessel, the piston will be driven into the tube with a force proportionate to its diameter; the atmosphere or air that surrounds it pressing with a weight equal to about fifteen pounds on each inch of its entire surface. On the heat being again applied, the process may be repeated with a similar result.

If the glass tube be lengthened, and bent in the form of an inverted U, or syphon, with the lower leg immersed in an open reservoir of water, thirty feet below the heated bulb, it will be found after a repetition of the process of condensation, that the pressure of the atmosphere, acting upon the surface of the water, will so far tend to fill up the vacuum, as to raise the water contained in the open reservoir to the top of the vessel; and it is upon this latter principle, that Savery's first engines were constructed; the remaining lift being effected by the repellent force of steam. As, however, we propose in this part of the work, to confine ourselves more particularly to the historical data, connected with the early history and invention of the steam engine, without reference to the more immediate principles of its construction and application, it may be enough to say, that an engine for raising water on this plan, has lately been contrived by Mr. Pontifex, of Shoe Lane, which for economy in the use of fuel, and in expense of erection, offers very considerable advantages over any yet employed.

In 1698, Captain Savery obtained a patent for a new mode of raising water, and communicating motion to a variety of machines by the force of steam; and in the following year a working model of the above engine, was submitted to the Royal Society, who then held their sittings in Arundel House. Savery's engine as we have already stated, was employed to raise water to a given height by the pressure of the atmosphere, and then to force the fluid up the remaining elevation, by the power of steam acting on the surface.

In the engines constructed under the authority of Savery's patent, it was necessary for a labourer to be in constant attendance for the purpose of turning the cocks, which alternately admitted steam and the condensing water. M. de Moura however effected a considerable improvement in this part of the engine, by constructing a self-acting apparatus for this purpose.

One of the greatest objections to this engine, was the extreme danger attendant on raising

---

* The following notice of this machine is inserted in their Transactions for that year.

" Mr. Savery, June 14, 1699, entertained the Royal Society, with shewing a small model of his engine for raising water by the help of fire, which he set to work before them. The experiment succeeded according to expectation, and to their satisfaction.'

water to any height beyond the atmospheric pressure; and as this seldom exceeded thirty feet, it was necessary for the remaining perpendicular lift to be effected by the expansive force of steam, upon the same principle as the Marquis of Worcester's engine, and for every thirty-three or four feet beyond that height, a pressure equal to the atmosphere must be exerted on the inside of the boiler and receivers, tending to burst them open. On this account it would require a separate engine, for every fourteen fathoms of the depth of a mine, thus raising the water from one to another; but in the event of any one becoming deranged, the whole must necessarily stop.

From this it will be seen, that this engine was not adapted either for the supply of towns or the draining of mines, (two of the patentee's principal objects,) the latter of which were often of considerable depth; but a number of small ones were erected for the raising of water in gentlemen's pleasure grounds, in different parts of England. Dr. Desaguliers tells us that he made seven of these engines: the first was for the Czar, Peter the Great, for his garden at Petersburgh, where it was set up. The boiler was made spherical, and held between five and six hogsheads. The receiver held one hogshead, and was filled and emptied four times in a minute. The water was drawn up by suction, or the pressure of the atmosphere, twenty-nine feet high out of the well, and then pressed up eleven feet higher. The pipes were all of copper, but soldered to the suction piece with soft solder, which held very well for that height. Had, however, the amount of pressure been greater, it must have burst the metal, and produced the most mischievous effects.

From a MS. in the possession of the late Rev. Mr. Stebbing Shaw, it appeared that Savery erected one of these about the close of the seventeenth century, at Wednesbury, in the county of Stafford. In this place the water rose so rapidly and in such quantities, from an adjacent coal pit, that it covered several acres of ground, where it remained for several years, although every effort was made to reduce it, and drain the land.

It was soon found that the mode of producing a vacuum in these engines was liable to the most serious objections, not the least of which was the unnecessary waste of steam; and an improvement on the original construction was shortly effected, which, although not capable of completely correcting this defect, produced a considerable saving in the amount of fuel employed.

It has been already stated, that Savery's mode of condensation consisted in throwing a quantity of cold water on the outside of the steam vessel employed to form a vacuum in the pipes; and on the re-admission of steam it was found necessary to restore the heat previously absorbed. To prevent this, a small jet was inserted, which striking against the steam, converted it into water without sensibly lowering the temperature of the vessel. When the repellent force of steam was employed in this engine, the waste of fuel was still more considerable; great part of the steam being condensed upon coming in contact with the surface of the water, so that it could not be brought into action till a large portion of the cold fluid was raised to the boiling point. As a further improvement on this engine, Papin introduced a moveable disk or float, which was interposed between the water and steam, and, by being pressed upon the former, forced it up the connecting pipe without the steam coming into actual contact with the water.

Newcomen, who is generally considered as the inventor of the atmospheric engine, appears to have been an ironmonger, resident at Dartmouth, in Devonshire; and that he was a man of considerable practical ingenuity, is sufficiently evident from his arrangement of the engine suggested by Dr. Papin.

Savery's engine having failed, from the causes we have already stated, the mines were nearly all at a stand for want of some cheap and efficient machine for the purpose of clearing the more distant workings. About this period Newcomen, having associated himself with John Cawley, a native of the same town, proposed to erect engines capable of supplying this desideratum, and taking the exhausted cylinder of Otto Guericke for a model, applied Papin's mode of producing a vacuum to the above machine.

To understand the action of this machine, we must conceive a hollow tube or cylinder, furnished with a piston, made to fit air-tight, and indeed in all respects similar to a common syringe. At the bottom of this are several apertures: one to communicate with the steam boiler, and furnished with a cock to open and shut the communication at pleasure; another for the admission of cold water; and a third to carry off the condensed steam and ejection water. A small lateral aperture is also formed with a valve to allow the escape of the air, or permanently elastic gas, which will not condense by the application of cold water: this last is called the *snifting clack*.

If the piston be now raised to the top of the cylinder, and steam admitted, the air will be ejected by the snifting clack. The steam is then cut off, and the cold water allowed to enter, which condensing, the steam forms a vacuum beneath the piston, which is pressed down with a force proportionate to its diameter.

In a working engine for the draining of mines, the piston rod is attached by a chain to the end of a long lever, working on a fulcrum, at the opposite end of which are suspended the rods of the pumps intended to raise the water: the weight of these rods exceeds the weight of the piston so much as to draw it up to the top of the cylinder, and the machine is thus ready for the admission of steam, and the production of an entire stroke.

The first really effective engine on this construction appears from a MS. to which we have already referred, to have been erected at Wolverhampton, near the half mile-stone leading from Walsingham to that town.

In 1718, the patentees agreed to erect an engine for the owners of a colliery, in the county of Durham, where several hundred horses were employed. Mr. Henry Beighton, who was engaged as an agent in this concern, not approving of the intricate manner of opening and shutting the cocks, for the admission of steam, water, &c. which were then moved by strings and catches, invented by a boy of the name of Potter, employed a hanging bar attached to the great working beam for that purpose.

But the most important improvement in the atmospheric engine, was the application of it by means of a crank and fly-wheel to the purpose of turning mills. It was first described by Mr. Hulls, who, in 1736, had a patent for an engine to be employed in the impelling of vessels. It was not, however, generally used till 1770, about which period it was attached to an engine, at Birmingham. Several other patents have also been obtained for this purpose; and, in 1781, the Abbé

Arnal, Canon of Alais, in Languedoc, entertained a thought of the same kind, proposing to employ it in inland navigation.

The extensive foundery of Colebrook-dale, appears to have had the honour of casting and finishing the first atmospheric engine of considerable size. It was intended for the use of Walker colliery, near Newcastle, where it was erected in 1763. The steam cylinder of this colossal engine weighed, exclusive of the bottom, 130 cwt. while the diameter of the bore, which was turned perfectly true, and well polished, measured upwards of seventy four inches. When completed, they were enabled to raise on an average above 307 cwt. of water at each stroke of the piston.

The double acting steam engine does not differ very materially from those we have already described. It was first suggested by Dr. Falck, who published an account of his invention in 1779. The chief improvement which he introduced was the use of two cylinders, into which the steam was alternately admitted by a common regulator, opening the communication with the steam to one, whilst it shut up the opening of the other. The piston rods were kept (by means of a wheel fixed to an arbor) in a continual ascending and descending motion, in the same manner as the rods of a common air-pump, while the nut, acting in the upright racks, was made to work the pumps, which were thus kept in constant action.

From this it will be seen, that in a double cylinder engine, where two cranks are used, the fly-wheel, which is usually employed as a magazine of power, may be entirely dispensed with; which, in the reciprocating engine, is an advantage of considerable importance, as the whole power of the engine must, in certain positions of the crank, depend upon the action of the fly-wheel.

The fire wheel of M. Amontons, and the steam wheel of his countryman Deslandes, were very ingenious, though both of them much too intricate for general use. The first of these inventions consisted of a number of buckets placed in the circumference of a wheel, and communicating with each other by very intricate circuitous passages. One part of this circumference was exposed to the heat of a furnace, and another to a cistern of cold water. The communications were so disposed, that the steam produced in the buckets on one side of the wheel, drove the water into buckets on the other side, so that one side of the wheel was always heavier than the other, and this constant addition of weight produced a rotatory motion.

Various attempts have been made at different periods to employ the steam engine in the draining of land. M. François was, we believe, the first who suggested its practical application to this purpose. He proposed to employ an engine on Savery's plan, and added machinery to open and shut the cocks. Two or three large engines have been constructed in this country, which have since been employed in Holland with the most beneficial effects; and there is no doubt but that their value, when duly appreciated, will be sufficiently obvious. This is more particularly the case in those tracts of low and swampy ground, whose outfall lies at a considerable distance, and which has previously to pass through ground of a higher level. In some instances it has been found necessary to cut drains or rather trenches of from ten to twenty feet in depth, and this too for several miles in length.

Mr. Savory, of Downham, who, we understand, has given considerable attention to this branch of civil engineering, states the cost of an engine of twenty-horse power fitted up for this purpose at fifteen hundred pounds, and that this will do as much work as a mill with a forty feet sail, when in full velocity. The advantages that may be derived from the use of steam in the fens or marsh country, appears, from the same authority, to be of the first importance. In case of intense frost, the uniform velocity, with the opportunities of communicating heat, would prevent the engine from freezing, to which, from the uncertainty of winds, the other engines are very much subject. The consequence is, that a great fall of snow coming at the same time that the mills have not been in a state to prepare the ditches to receive the over-plus water which it occasions, an inundation generally takes place in the fens; and, as the waters rise very rapidly under these circumstances after a thaw, it frequently occurs, that when the mills are set at liberty from the effects of ice, they are for some days incapable of successfully opposing the accumulation of water. On the other hand, by adopting the means of steam, the engines would be working in full effect during the continuance of a frost, and the ditches being kept proportionably low, would at all times be capable of discharging the water, and thus prevent inundation.

Mr. S. concludes this part of a very useful paper on the subject, by observing, that "as to a district of country which requires draining without any engines upon it, at the time of its being undertaken, it is a matter of doubt in my mind, whether it could not be drained more economically by steam, than by the means usually adopted, although the expence of fuel must certainly be very great. Taking the average of winds, the mills in the winter season do not throw so much water in a week, as they would in one third of the time, if they went with all the velocity of which they are capable. It follows, that one steam engine, with equal powers, would do as much execution in the course of a season, as three windmills, and consequently a great saving would accrue in the first expense, and afterwards in attendance and repairs."*

* Vide Communications to the Board of Agriculture, vol. iv. p. 52.

IMPROVEMENTS

EFFECTED BY

MR. WATT, AND OTHERS,

DOWN TO THE PRESENT TIME.

CHAP. II.

Boulton and Watt—Cartwright—Smeaton—Hornblower — High-pressure Engine — Wolf's Improvements—Rotatory Engines—Kempel — Sadler — Cooke — Bell-crank Engine — Employment of the Steam Engine in North America, and the Colonies—Locomotive Engines.

In the engine usually ascribed to Newcomen, the steam was not employed as an impelling power, but was used for producing a vacuum beneath the piston, which was afterwards forced down by the pressure of the atmosphere; and it was left to the masterly and towering genius of an otherwise obscure mechanic, to quadruple the force of this stupendous machine, and by one step, perfect the labours of the preceding century.

Mr. Watt's attention was first drawn to this subject, by an examination of a small model of an atmospheric engine, belonging to the University of Glasgow, which he had undertaken to repair. In the course of his experiments with it, he found the quantity of fuel and injection water it required, much greater in proportion than in the larger engines; and it occurred to him, that this must be owing to the cylinder of this small model exposing a greater surface in proportion to its contents, than was effected by larger cylinders. This he endeavoured to remedy, by employing non-conducting substances for those parts of the engine which came in immediate contact with the steam. After a variety of experiments, the results of which we shall presently describe, he succeeded in constructing a working model, capable of producing a force equal to fourteen pounds on every inch of the piston, and which did not require more than one third of the steam used in the common atmospheric engine to produce the same effect.

It will be evident that this was as near an approximation towards perfection as could possibly have been expected; and indeed much more than was likely to be effected in a large engine, as the vapour left beneath the piston possessed only one-fifteenth part of the elastic force of the steam employed to form the vacuum.

Having discovered that the great waste of caloric in the old engine, arose from the alternate heating and cooling the cylinder, by the admission and subsequent condensation of the heated steam, Mr. Watt perceived that to make an engine in which the destruction of steam should be the least possible, and the vacuum the most perfect, it was necessary that the cylinder should remain uniformly at the boiling point; while the water forming the steam was cooled down to the temperature of the atmosphere. To effect this, he employed a separate condensing vessel, between which, and the hot cylinder, a communication was formed by means of a pipe and stopcock.

To understand the action of this engine, we may employ a common syringe, connected with a boiler, as in the atmospheric engine, and furnished with a pipe passing into an air-tight vessel, immersed in water for the purpose of condensation.

If the piston be then raised, and the communication with the condenser cut off, the steam will speedily expel the air; when this is effected, the further admission of steam must be prevented, and the communication with the condenser opened. The steam will now expand itself, passing down the pipe and entering the condenser; the moment, however, that it comes in contact with the sides of the cold vessel, it will be condensed and a vacuum formed; and this process will continue to proceed, so long as any steam remains beneath the piston.

The only objection that offered itself to this admirable mode of condensation, arose from the difficulty experienced in getting rid of the water and air that remained in the condensing vessel. When steam was generated from water that had been freed from air by long boiling, a considerable advantage was obtained; and it was found that a power nearly equal to the entire pressure of the atmosphere was produced. The great advantage thus obtained will be sufficiently obvious, when it is known that, in the engines previously constructed, the elasticity of the steam arising from the heated injection water remaining at the bottom of the cylinder, was equal to one-eighth of the atmospherical pressure, and consequently destroyed an equal proportion of the power of the engine.

The mode of condensing the steam, by the application of cold water to the outside of the condenser, was soon found inconvenient from the great size and expense attendant on the use of this apparatus; and Mr. Watt introduced an internal jet of cold water, which, striking against the steam, instantaneously reduced it to its original bulk, and thus formed a vacuum. To draw off the condensing water, as well as to get rid of the air that was extricated during condensation, he found it necessary to employ a small pump, worked by the engine, the size of which was proportioned to the amount of air and water generated in the condenser. In one of the early engines upon this construction, erected at Bedworth, three air-pumps were used; two below, worked by chains connected with the beam, and a third, placed above, which received the hot water raised by the others. In the engines now constructed, only one air-pump is employed, and this fully answers the intended purpose.

Another improvement introduced by Mr. Watt, consisted in surrounding the upper part of the cylinder with a cap, through a hole in the centre of which the piston rod worked air-tight. The force of steam was then substituted for that of the atmosphere, and at a pressure of more than fifteen pounds on the square inch; so that when a vacuum was formed beneath the piston, steam of considerable impellent force was entering the upper end of the cylinder, by means of a pipe connected with the boiler.

By thus substituting the force of highly elastic vapour, for the ordinary pressure of the atmosphere, the upper and under side of the piston were preserved at the same temperature, and the supply of steam being regulated by the width of the aperture, any given amount of force might readily be produced. In the atmospheric engine this could not be effected, as the whole pressure of the atmosphere was made to act on the piston, the instant the vacuum was formed by the condensation of the vapour beneath; so that in the event of a pump-rod breaking, by which the elevation of the water might be impeded, and the labour of the engine taken off, the rapid descent of the piston would evidently cause the destruction of the entire apparatus.

Soon after the completion of his first model, Mr. Watt erected an engine for his friend Dr. Roebuck of Kinneil, near Borrowstownness, with whom he was afterwards associated in the manufacture of his improved engine: the latter gentleman, however, in 1774, disposed of his share of the business to Mr. Boulton, of Soho*.

---

* From this nursery of ingenuity has originated some of the noblest and most striking *chefs d'œuvre* of mechanical art yet witnessed. The following account of this celebrated manufactory is from the pen of Dr. Darwin. It was written in 1768, and when the manufactory, although " big with promise," was little more than a type of its present magnitude.

" Soho is the name of a hill in the county of Stafford, about two miles from Birmingham, which a very few years ago was a barren heath, on the bleak summit of which stood a naked hut, the habitation of a warrener.

" The transformation of this place is a recent monument of the effects of trade on population. A beautiful garden, with wood, lawn, and water, now covers one side of this hill; five spacious squares of building, erected on the other side, supply workshops or houses for about six hundred people. The extensive pool at the approach to this building, is conveyed to a large water-wheel in one of the courts, and communicates motion to a prodigious number of different tools. The mechanical inventions for this purpose are superior in multitude, variety, and simplicity, to those of any manufactory in the known world.

" Toys, and utensils of various kinds, in gold, silver, steel, copper, tortoise-shell, enamels, and many vitreous and metallic compositions, with gilded, plated, and inlaid works, are wrought up to the highest elegance of taste, and perfection of execution in this place.

" Mr. Boulton, who has established this great work, has joined taste and philosophy with manufacture and commerce; and from the various branches of chemistry, and the numerous mechanic arts he employs, and his extensive correspondence to every corner of the world, is furnished with the highest entertainment, as well as the most lucrative employment.

We have already stated, that Dr. Falck described an engine with two cylinders as early as 1779, which he called a *double acting engine;* but similar advantages were obtained by Mr. Watt in an engine with a single cylinder. To effect this, he applied the power of the steam to press the piston upwards as well as downwards, by forming the vacuum alternately above and below the cylinder. When it became necessary to connect the piston and beam for an upward, as well as for a downward stroke, a double chain acting on an arch head was substituted for the single one previously employed; and this was speedily superseded by a rod connected with the working beam, by means of a parallel motion.

The *Expansion Engine* was also invented by Mr. Watt, and though not generally employed until 1778, appears from a letter written by him to a gentleman of Birmingham, to have suggested itself as early as 1769. The principle of this invention, consists in shutting off the farther entrance of steam from the boiler when the piston has been pressed down in the cylinder, for a certain proportion of its total descent, leaving the remainder to be accomplished by the expansive force of that steam already introduced.

To regulate the time of closing this valve, and as such the precise amount of steam admitted, Mr. Watt employs a plug-frame with moveable pins, which may be placed in such a manner, that the steam-valve will shut when the piston has descended one-half, one-third, one-fourth, or at any other proportion. By the application of this principle the piston is made to descend with an uniform velocity, the pressure on the piston continually diminishing as the steam becomes more and more rare, and the accelerating force which works the engine is consequently diminished.

The advantages attendant on this mode of admitting steam, are however greater in an high-pressure engine, than in those usually constructed by Messrs. Boulton and Watt.

It does not appear that Mr. Watt produced any large engines on his improved construction until 1774; and so slow is the progress of improvement against preconceived habits of long standing, or interested clamour, that he found the term of his patent was likely to pass away before he should be reimbursed. In consequence of this he applied to parliament, which, by a legislative enactment, sanctioned the prolongation of the original term for twenty-one years.

Soon after the renewal of his patent, Mr. Watt published proposals for the erection of his improved engine, and in these the advantages to be obtained from its use are placed in the strongest light. In an atmospheric engine, constructed on the most improved plan, a quantity of water equal to 9,636,660 pounds, was raised one foot high with a single bushel of Newcastle coals; but Mr. Watt undertook to raise 24,553,571 pounds, with the same amount of fuel.

The increasing depth of some of the larger works in the mining districts, has called for a proportionate increase of the power employed; and many of the engines now erected for the purpose of raising water, are of proportions so gigantic, as can be conceived only by actual calculation. One of the largest is that erected at the Union mine in Cornwall. It is upon Mr. Watt's double-action principle, and is loaded to exert its utmost force. The steam-cylinder is sixty-three inches diameter, and raises water equally in the ascent or descent of the piston. The weight of water in the pumps is 82,000 pounds; and with this load it makes 6½ double strokes per minute, of 7¾ each, or it gives to the load 100¾ feet motion per minute. To effect this, which is equivalent to raising 8,261,500 pounds, lifted one foot high, or the power of 250 horses, it is necessary to consume about thirty-one pounds of coal per minute.

In 1797, the ingenious Mr. Cartwright, well known for the value and variety of his scientific avocations, invented a mode of applying the vapour of alcohol, or other ardent spirit, as a substitute for common steam. In addition to the saving to be effected by this plan, Mr. Cartwright intended to employ his engine as a still, by which the whole cost of fuel would be saved*. Such a

* It is curious, that a distinguished chemist of the present day should have suggested the construction of an engine on the following plan, which, it will be seen, is precisely similar to that described in Mr. Cartwright's specification.

method, however, supposes a capability of blending the business of a distiller with a variety of trades, to which it is totally inapplicable. A scheme somewhat similar to this, and to which we shall afterwards more fully revert, has lately been attempted by Colonel Congreve, in which he proposes to burn a large portion of chalk mixed with the coal, and thus convert the furnace into an

---

" Since the vapour of alcohol, having the same elastic force as the atmosphere, contains $\frac{44}{100}$ of the latent heat of ordinary steam, and since its elastic force is doubled at the 206th degree (six below the boiling heat of water) with perhaps one-third of additional caloric, might we not, in particular circumstances, employ this vapour for impelling the piston of a steam engine? The condensing apparatus, could be, I imagine, so constructed, as to prevent any material loss of the liquid, while more than a quadruple power would be obtained from the same size of cylinder at 212°, with an expenditure of fuel not amounting to one half of what aqueous vapour consumes; or the power and fuel would be as three to one, calling their relation in ordinary steam one to one. A considerable engine could thus also be brought within a very moderate compass. Possibly, after a few operations of the air pump, the incondensible gas may be so effectually withdrawn, that we might be permitted to detach this mechanism, which, though essential to common engines, takes away one-fourth of their power. In a distillery in this country, or on a sugar estate in the colonies, a trial of this plan might perhaps be made with advantage. While exercising its mechanical functions of grinding, mashing, or squeezing the canes, it would be converting ordinary into strong spirit for rectification, or for the convenience of carriage. Might not such an engine be erected on a small scale, for many purposes of domestic drudgery? It would unquestionably furnish a beautiful illustration in philosophy, to make one small portion of liquid, by the agency of fire, imitate the ceaseless circulation and restless activity of life."—Phil. Tran. vol. cviii. p. 393.

efficient lime-kiln. From this view of the subject, we think it will be seen, that however plausible or ingenious this invention may appear in theory, there are insuperable objections to its general employment. We are still, however, greatly indebted to Mr. Cartwright for the mechanical arrangement of the engine, described in this patent, as it furnishes the first hint for an elastic metal piston, which has since been found of the greatest use in high-pressure engines.

The first Portable Steam Engine appears to have been constructed by Mr. Smeaton, who employed it for draining foundations and other temporary works. It had a pulley, or wheel, to receive the chain which communicated motion from the piston to the pump-rod, instead of a beam. The pivots of the wheel were supported by two inclined beams connected at top, whilst the cylinder and pump were bolted down to the groundsills. Thus, the whole machine being supported by one frame of wood, it could without trouble be set to work in the open air. The boiler, which required no setting in brick work, was in the shape of a tea-kettle, and the fire-place being in the centre, was surrounded on all sides by water, thus presenting the greatest possible surface to the action of the flame. Portable steam engines are now employed not only in the erection of bridges, and in underground excavations, but are also usefully applied to the purpose of propelling vessels and carriages: the latter application is of a very recent date. Steam navigation, however, from its great national importance, will deservedly find a place in a separate division of this work.

Of Mr. Hornblower's engine, but little need be said, as in practice he was scarcely able to obtain a greater effect from the expansive action of the steam in two cylinders, than Mr. Watt's did in one; and having attempted in vain to procure an extension of the original term of his patent, he was afterwards prosecuted by Messrs. Boulton and Watt for an infringement on theirs, in using the condenser and air-pump.

The principle of the high-pressure steam engine depends on the power of steam to expand itself very considerably beyond its original bulk, by the addition of a given portion of caloric, thus acquiring a considerable elastic force, which, in this case, is employed to give motion to a piston. One of the greatest advantages attendant on employing the repellent force of steam, as in this form of the engine, consists in an evident saving of the water usually employed in condensation; and this, in locomotive engines, for propelling carriages, is an object of considerable importance.

Leupold has furnished a description of a high-pressure engine, in a very valuable work on machines, published in 1724. He ascribes the invention to Papin, to whom we have already alluded as the inventor of the atmospheric engine. The apparatus described by Leupold, consists of two single cylinders, placed at some distance from each other, each of which is provided with a piston made to fit air-tight, and connected with a forcing pump.

When steam of considerable elasticity is admitted at the bottom of the first cylinder, it is forced upwards, carrying with it the lever of the pump; at the same time that the steam or air is expelled from the other. On this operation being repeated, or rather reversed, the steam is allowed to enter the second cylinder, which is also connected with the boiler, while the steam in the first cylinder is allowed to escape into the air. From this it will be evident that the process of condensation forms no part of the principle of the high-pressure engine; and that the expansion of gunpowder might be made to produce a precisely similar effect.

The amazing force to be produced by the expansion of highly elastic vapour, did not escape the penetrating notice of that towering genius, which was now directing all its energies towards its improvement. Accordingly, we find in Mr. Watt's first patent, the following clause, which expressly describes this engine: " I intend, in many cases, to employ the expansive force of steam to press on the pistons, or whatever may be used instead of them, in the same manner as the pressure of the atmosphere is now employed in common fire-engines. In cases where cold water cannot be had in plenty, the engines may be wrought by the force of steam only, by discharging the steam into the open air after it has done its office."

Messrs. Trevithick and Vivian were the first to employ the high-pressure engine to advantage, as they found it admirably adapted for the purpose of propelling carriages. In this case the steam, after having performed its office, was thrown off into the air; and the condenser, together with the necessary supply of cold water which must have accompanied it, was by this means dispensed with. For the purpose of motion, the high-pressure engine certainly possesses considerable advantages, not the least of which is its cheapness and portability; the danger, however, attendant on the use of steam, acting with a force equal to from forty to eighty pounds on the square inch, must inevitably form an insuperable bar to its general introduction to our manufactures.*

Mr. Woolf's improvements, which are of considerable importance, are founded on the same principle as those of Mr. Watt, namely, the power

---

* To understand the power of a high-pressure engine, and the amount of caloric absorbed to produce a given force of steam, it is necessary to compute this force at the various thermometine degrees of heat, commencing at the boiling point of water. This has been most accurately calculated by Dr. Ure, who, as well as Mr. Dalton, made a variety of experiments to illustrate this highly interesting subject, the result of which will be found in Appendix (B).

of steam to expand itself, or increase its volume in a very considerable degree, after its passage from the boiler. From a variety of experiments made on this subject, he ascertained that a quantity of steam having the force of five, six, seven, or more pounds on every square inch of the boiler, may be allowed to expand itself to an equal number of times of its own volume, when it would still be equal to the weight of the atmosphere, provided that the cylinder in which the expansion takes place, have the same temperature as the steam possessed before it began to increase.

The increased degree of temperature necessary, both to produce and to maintain this augmentation of the steam's expansive force is considerable; and the amount may be seen from the following table:

	Pounds per square inch.		Degrees of heat.		Expansibility.	
Steam of an elastic force sufficient to overcome the pressure of the atmosphere upon a safety-valve	5	requires to be maintained by a temperature equal to about	$227\frac{1}{2}$	and at these respective degrees of heat, steam can expand itself to about	5	times its volume, and continues equal in elasticity to the pressure of the atmosphere.
	6		$230\frac{1}{4}$		6	
	7		$232\frac{3}{4}$		7	
	8		$235\frac{1}{4}$		8	
	9		$237\frac{1}{2}$		9	
	10		$239\frac{1}{2}$		10	
	15		$250\frac{1}{2}$		15	
	20		$259\frac{1}{2}$		20	
	25		267		25	
	30		273		30	
	35		278		35	
	40		282		40	

The most economical mode of employing this principle, consists in the application of two steam-cylinders and pistons of unequal size to an high-pressure boiler; the smaller of which should have a communication both at its top and bottom, with the steam vessel. A communication being also formed between the top of the smaller cylinder and the bottom of the larger cylinder, and *vice versa.* When the engine is set to work, steam of a high temperature is admitted from the boiler to act by its elastic force on one side of the smaller piston, while the steam which had last moved it, has a communication with the larger or condensing cylinder. If both pistons be placed at the top of their respective cylinders, and steam of a pressure equal to forty pounds the square inch, be admitted, the smaller piston will be pressed down, while the steam below it, instead of being allowed to escape into the atmosphere, or pass into the condensing vessel, as in the common engine, is made to enter the larger cylinder above its piston, which will make its downward stroke at the same time as that in the smaller cylinder; and, during this process, the steam which last filled the larger cylinder, will be passing into the condenser to form a vacuum during the downward stroke.

To perform the upward stroke, it is merely necessary to reverse the action of the respective cylinders; and it will be effected by the pressure of the steam in the top of the small cylinder, acting beneath the piston in the great cylinder; thus alternately admitting the steam to the different sides of the smaller piston, while the steam last admitted into the smaller cylinder, passes regularly to the different sides of the larger piston, the communication between the condenser and steam boiler being reversed at each stroke.

The economical application of this engine, may however be best understood by an examination of its effective force when applied to the raising of water. It appears that a double cylinder expansion engine was constructed for Wheal Vor mine in 1815. This has a great cylinder of fifty-three inches in diameter, and nine feet stroke, the small cylinder being about one-fifth of the contents of the great one. The engine works six pumps, which at every stroke raise a load of water of 37,982 lbs. weight, seven feet and a half high. This produces a pressure of 14.1 lbs. per square inch on the surface of the great piston, while its average performance may be estimated at 46,000,000 lbs. raised one foot high with each bushel of fuel.

The great mass of inert matter contained in the working beam of the reciprocating engine, must of necessity produce a proportionate waste of power; each elevation of the piston causing a change from a state of rest to motion, and *vice versa.* This, however, is in no small degree enhanced, by the necessity of employing a fly-wheel of considerable weight to equalise its motion. To prevent this loss of power, a variety of contrivances have been suggested, for the purpose of producing a continuous action, without the intervention of a cylinder and piston, thus dispensing with the beam and fly-wheel.

To produce a continuous rotatory motion, is, however, little more than a return to the simple principles on which the earliest engines were constructed. We have already stated, that the Italian philosopher Brancas, directed steam of considerable repellent force against the vanes of a wheel, which was employed to give motion to a stamping press, though the force obtained must have been very inconsiderable. The principle of this invention was also applied to a very ingenious, though we fear useless rotatory engine, constructed by Kempel.

It consisted of a hollow cylinder, furnished with two arms, at the end of which were transverse apertures; and this was connected with a large æolipile or boiler, by means of a moveable socket. The steam employed to drive the arms was of considerable elasticity, and rushing out at the apertures with considerable violence, produced a rotatory motion.

About the same period, Mr. Sadler, of Oxford, took out a patent for a similar machine, though we are not aware that it was ever usefully employed, the rarity of steam being such, that even if none be condensed by the cold of the atmosphere, the impulse must necessarily be exceedingly feeble, and the expense of steam, to produce any serviceable effect on the machinery, exceedingly great.

A very ingenious, and not less simple mode of producing a continuous rotatory motion, is described by Mr. J. Cooke in the Transactions of the Irish Academy.* It consisted of a wheel, with moveable valves or flaps on its circumference, turning freely on joints, which admitted their openings only to a line passing through the centre of the wheel. These, when closed, lay exactly on a level with its outer extremity, but when open fell down perpendicular to it. The wheel, thus formed, was enclosed in a case, which fitted it exactly, while the upper valves were close and the lower extended. The steam in its passage from the boiler to the condenser, pushed the extended valves, and thus gave motion to the wheel. A working model of this engine, without the condenser, was, we believe, exhibited before several members of the academy. In this instance, however, it must have acted as a high-pressure engine, discharging the steam at each division of the wheel; and we very much doubt whether it would be possible to pack the hanging valves sufficiently tight to admit of their readily falling to the position described by Mr. Cooke.

Several other attempts have also been made to produce a continuous rotatory motion, the most important of which will be found in the analysis of patents prefixed to this work; but the only one which appears at all likely to succeed in practice has lately been invented by Mr. Masterman, and to this we would more particularly call the

* Vol. iii. p. 113. A.D. 1789.

reader's attention. In his engine, Mr. Masterman proposes to employ water as the propelling power, which he effects, by enclosing it in the tubular rim of a large wheel, furnished with valves opening in one direction. This wheel is made to revolve on a hollow axis connected with the steam boiler. The arms which radiate from the axis are also hollow; and on the admission of steam from the boiler, it is conducted through the arm immediately opposite, and entering the rim of the wheel, comes in contact with, and presses against the column of water on the one side, and the closed valve on the other. The water being previously heated to the boiling point, no condensation ensues, but the whole weight of water, which was previously balanced in two columns of equal length, is driven by the pressure of the steam to the side opposite to that at which the elastic vapour entered, and that side of the wheel will necessarily preponderate. If this process be repeated, the steam being allowed to blow through each radiating arm in succession, a continuous rotatory motion will be produced. Should it be found adviseable to employ steam of less elasticity, a condenser may be added, and this too without materially increasing the expense.

In Messrs. Boulton and Watt's *Bell-crank* Engine, the cylinder is supported by brackets proceeding from a cast-iron condensing-cistern, and is placed over one end of it. The beam which is formed like a right-angled triangle, has its centre of motion at the right angle, and the axis of it is supported by bearings screwed to the cistern; and at the opposite end to that upon which the cylinder is placed, the horizontal arm of the triangle forms the working arm of the beam, to the extremity of which the power of the cylinder is applied. The connecting rod is jointed at the upper end of the perpendicular arm, and extends to the crank, which is supported in bearings screwed to the cistern at the same end at which the cylinder is placed, the centre of motion being at the same level with the top of the cistern; and beneath the cylinder, the hypothenuse of the triangle of the beam forms a brace to strengthen it. Two of these beams are used, and are applied on opposite sides of the cistern, upon the same axis of motion, and are united together by cross rods, thus forming two connecting rods and cranks upon one axis of motion, the fly-wheel being placed at one extremity of the axis. To connect the piston-rod with the ends of the arms of the beam, or the base of the triangle, a rod is fixed across the top of the piston rod; and to the two ends of this two other rods are linked, which descend to the beam, and are pointed to it at the ends. By this means, the ascent and descent of the piston-rod produce a corresponding action of the beam upon its centre of motion; the upper end of the perpendicular arm moving backwards and forwards, thus by means of the connecting rods turning the cranks. The rods which descend from the bar, which is fixed across the top of the piston-rod to the ends of the beams, are so constructed, as to preclude the necessity of employing the parallel motion. This engine is very compact; it requires no fixing, and the air-pump which is placed in the middle of the cistern, is worked by two rods jointed to the horizontal arms of the beams.

The atmospheric engine was first employed in North America, about 1760, and two engines on this principle were erected in New England before the revolutionary war. One was also employed for the purpose of draining Schuyler copper mine, on the river Passaick, in New Jersey. The greater part of these engines, however, were executed in England. Some idea of the slow progress which has been made by our transatlantic brethren, in this very important invention, may be furnished by the following fact, that no more than four engines of any importance were at work in the whole continent of America at the beginning of the present century. Of these, two were erected at New York, the first of which was employed for the supply of that city with water,* and the other gave motion to a saw-mill. The remaining two were erected at Philadelphia, and belonged to the corporation of that city.

---

* This engine is upon the principle of Boulton and Watt's double action engines. It has two boilers; one of wood, and the other of sheet iron. The fly is driven by a sun-and-planet wheel, and the shaft works three small pumps.

The city of Philadelphia was for many years supplied with water by means of a low-pressure engine; Mr. Evans, the Trevithick of America, has, however, lately substituted a high-pressure engine in its place. By this means, in fuel alone, a saving of seventeen dollars per day has been effected.

Trinidad appears to have been the first West Indian colony in which the resident planters could overcome their ancient prejudice in favour of the cattle mills, which are still generally employed on sugar estates. Mr. William Lushington, who had a considerable property on that island, introduced a large engine by Boulton and Watt in 1804; and so great was the saving effected by its use, that the labour of three mills, each of which were equally expensive, was usually performed by the one machine.

From the result of some experiments lately made with two steam engines, constructed by Mr. Maudslay for the island of Ceylon, it appears, that this species of power has been equally efficient in the dressing of rice and preparation of oil. The apparatus employed for this purpose was contrived by Mr. Hoblyn, a gentleman well known for his distinguished scientific attainments, and it has been found that, allowing three hundred working days in the year, and the engines to work ten hours per day, they would annually convert 576,000 of paddy into a quantity of rice equal to 147,347 bags, which, at nine rix-dollars a bag, would amount to 116,035*l.* sterling, while the same quantity of paddy being converted into rice by the common method, would produce only 64,799*l.* The saving, therefore, by these two machines, is more than 50,000*l.* per annum.

But the introduction of this valuable mechanical agent to the mines of South America, did not take place till 1815; and in the following year a report on the subject was published in the Lima Gazette. After describing this important event, it adds, "Immense and incessant labour, and boundless expense, have conquered difficulties hitherto esteemed altogether insuperable; and we have, with unlimited admiration, witnessed the erection, and astonishing operation of the first steam-engine. It is established in the celebrated and royal mineral territory called the mountain Yaiiricoeha, in the province of Tarma; and we have the felicity of seeing the drain of the first shaft in the Santa Rosa mine, in the noble district of Pasco." They add, "We are ambitious of transmitting to posterity, the details of an undertaking of such prodigious magnitude, from which we anticipate a torrent of silver, that shall fill surrounding nations with astonishment."

It appears that the new world is principally indebted to the agency of M. François Uville for this improved era in their mining annals. This gentleman having found that a large portion of the most valuable mines in Peru were falling into decay, and in some cases totally drowned from the impossibility of draining them by manual labour, applied to Mr. Trevithick of Camborne in Cornwall, one of the patentees of the high-pressure engine. This ingenious mechanic applied himself with such extraordinary diligence to the subject, that in less than nine months the materials for as many engines were completely ready for their destination. This apparatus, which cost about ten thousand pounds, left Portsmouth in the beginning of September, 1814, accompanied by M. Uville and three Englishmen, to direct the erection of the machinery.

Mr. Trevithick was afterwards employed to superintend the Royal Mint established at Lima, and on his arrival in South America, was received with such enthusiastic gratitude, that the Lord Warden proposed to erect his statue in massive silver. The engines employed were exclusively on the high-pressure principle, and will be found under his patent in the Appendix to the present work. Indeed this appears to be the only engine likely to act with an advantageous effect, the extreme rarity of the atmosphere in those elevated regions, precluding the economical use of the common engine.

We have hitherto viewed the steam engine, when employed as a substitute for animal force, in giving motion to mills, raising of water, and a variety of other employments, all of which, however, are of a fixed and stationary nature. But some progress has likewise been made towards the application of the same power to moveable machinery, and when constructed for this purpose it is called a *locomotive engine*.

The employment of an internal mechanism to impel waggons on a plane road is of very early date, but the first application of the steam engine to this purpose took place, we believe, in the Royal Arsenal at Paris, towards the close of the last century. From this time till 1802, but little progress appears to have been made in the use of this species of wheel carriage; but about the latter period, Mr. Trevithick commenced a series of experiments on the use of the high-pressure engine for the above purpose; and this, with some improvements, has since been adopted.

When these engines were first tried, it was found difficult to produce a sufficient degree of re-action between the wheels and the track road, so that the former turned round without advancing the vehicle. This was remedied by Mr. Blenkinsop, who, when he adopted this species of conveyance, took up the common rails on one side of the whole length of the road, and replaced them with rails which had large and coarse cogs projecting from the outside. The impelling wheel of the engine was made to act in these teeth, so that it continued to work in a rack the whole length of the road.

An engine of four horses' power, employed by Mr. Blenkinsop, impelled a carriage lightly loaded at the rate of ten miles an hour, and when connected with thirty coal waggons, each weighing more than three tons, it went at about one-third of that pace.

The application of the steam engine to impel carriages on the public roads, has hitherto been considered as a refinement in mechanics, rather to be wished for, than a matter of reasonable expectation. It has however been stated, that a vehicle of this description is now constructing in Ireland, intended as a stage-coach, and it is added, that when loaded with a weight equal to four tons, it will be enabled to advance at the rate of fifteen English miles per hour. But it must, we think, be sufficiently apparent that the employment of this species of prime mover on a common gravel road, would be in the highest degree destructive, and a considerable increase in the toll would be the certain consequence.

In proof, however, that the necessity of employing an iron track-road for these vehicles is not so serious an objection as at first view might be supposed, more particularly in our mining districts, the neighbourhood of Newcastle alone, affords, within an extent of twenty-eight square miles, more than seventy-five miles fitted for this species of conveyance; and it is a well known fact, that there are many situations in which iron rail-roads might be advantageously employed, in which it would be quite impossible to open a navigable canal.

STEAM NAVIGATION.

CHAP. III.

Introduction and Improvements effected by Hulls—Duquet—Jouffroy—Fulton—Miller—Symington—Stanhope—Linnaker—Thames and Clyde boats—Progress of Steam Navigation in America.

THE possibility of employing steam as a moving power in the navigation of vessels, was known early in the last century; its practical application however, on a large scale, has not been fully established above twenty years.

In 1698, Savery recommended the use of paddle-wheels, similar to those now so generally employed in steam vessels, though without in the remotest degree alluding to his engine as a prime mover; and it is probable that he intended to employ the force of men or animals working at a winch for that purpose. About forty years after the publication of this mode of propelling vessels, Mr. Jonathan Hulls obtained a patent for a vessel in which the paddle-wheels were driven by an atmospheric engine of considerable power.

In describing his mode of producing a force sufficient for towing of vessels, and other purposes, the ingenious patentee says, " In some convenient part of the tow-boat there is placed a vessel about two-thirds full of water, with the top close shut; this vessel being kept boiling, rarefies the water into a steam; this steam being conveyed through a large pipe into a cylindrical vessel, and there condensed, makes a vacuum, which causes the weight of the atmosphere to press on this vessel, and so presses down a piston that is fitted into this cylindrical vessel, in the same manner as in Mr. Newcomen's engine, with which he raises water by fire.

"It has been already demonstrated that when the air is driven out of a vessel of thirty inches diameter, (which is but two feet and a half,) the atmosphere will press on it to the weight of 4 tons 16 cwt. and upwards; when proper instruments for this work are applied to it, it must drive a vessel with great force."

Mr. Hulls' patent is dated 1736, and he employed a crank to produce the rotatory motion of his paddle-wheels, and this ingenious mode of converting a reciprocating into a rotatory motion, was afterwards recommended by the Abbé Arnal, Canon of Alais in Languedoc, who, in 1781, proposed the crank for the purpose of turning paddle-wheels in the navigation of lighters.

It is probable that Mr. Hulls anticipated some objection to his new mode of propelling vessels, and it appears from Captain Savery's statement, to which we have already alluded, that a strong prejudice had been raised against the use of propelling wheels in vessels. Mr. Secretary Trenchard, who was at that time at the head of the Admiralty, had also given a decided negative to the proposition. In answer therefore to the objections which might have been anticipated, Mr. Hulls proposed the following queries, which he afterwards solved in the most satisfactory way.

"*Query* 1.—Is it possible to fix instruments of sufficient strength to move so prodigious a weight, as may be contained in a very large vessel?

Answer.—All mechanics will allow it is possible to make a machine to move an immense weight, if there is force enough to drive the same, for every member must be made in a proportionable strength to the intended work, and properly braced with laces of iron, so that no part can give way, or break.

Query 2.—Will not the force of the waves break any instrument to pieces that is placed to move in the water?

Answer. First, It cannot be supposed that this machine will be used in a storm or tempest at sea, when the waves are very raging; for if a merchant lieth in a harbour, &c. he would not choose to put out to sea in a storm, if it were possible to get out, but rather stay until it were abated. Secondly, when the wind comes ahead of the tow-boat, the fans will be protected by it from the violence of the waves, and when the wind comes side-ways, the waves will come edge-ways of the fans, and therefore strike them with the less force. Thirdly, there may be pieces of timber laid to swim on the surface of the water on each side of the fans, and so contrived as they shall not touch them, which will protect them from the force of the waves.

Up inland rivers where the bottom can possibly be reached, the fans may be taken out, and cranks placed at the hindmost axis to strike a shaft to the bottom of the river, which will drive the vessel forward with the greater force.

Query 3. It being a continual expense to keep this machine at work, will the expense be answered?

Answer. The work to be done by this machine will be upon particular occasions, when all other means yet found out are wholly insufficient. How often does a merchant wish that his ship were on the ocean, when, if he were there, the wind would serve tolerably well to carry him on his intended voyage, but does not serve at the same time to carry him out of the river, &c. he happens to be in, which a few hours' work at this machine would do. Besides, I know engines that are driven by the same power as this is, where materials for the purpose are dearer than in any navigable river in England. Experience, therefore, demonstrates, that the expense will be but a trifle to the value of the work performed by those sort of machines, which any person who knows the nature of those things may easily calculate."

M. Duquet appears to have tried revolving

oars as early as the year 1699, and experiments were made with them on a large scale both at Marseilles and at Havre:* this mode, however, of impelling vessels was soon given up as impracticable; and after our countryman, Hulls, the Marquis de Jouffroy unquestionably holds the most distinguished rank in the list of practical engineers, who have added to the value of this invention.

It is evident from an article published in the *Journal des Debats*, that in 1781 the Marquis constructed a steam-boat at Lyons, of 140 feet in length. With this he made several successful experiments on the Saone, near that city. The events of the revolution, which broke out a few years afterwards, prevented M. de Jouffroy from prosecuting this undertaking, or reaping any advantage from it. On his return to France after a long exile, in 1796, he learned from the newspapers that M. De Blanc, an artist of Trevoux, had obtained a patent for the construction of a steam-boat, built probably from such information as he could procure relative to the experiments of the Marquis. The latter appealed to the government, which was then too much occupied with public affairs to attend to those of individuals. Meanwhile Fulton, who had gained the same information, and was making similar experiments

---

* Vide Recueil de Machines approuvées par L' Académie Royale de Sciences, tome i. 173.

near the Isle des Cygnes, alarmed M. De Blanc, who knew that he had much more to fear from the influence and mechanical skill of an Anglo-American, than from that of an emigrant. He accordingly alleged his patent right, and requested the stoppage of Mr. Fulton's works, who returned for answer, that his essays could not affect France, as he had no intention to set up a practical competition upon the rivers of that country, but should soon return to America, which he actually did, and commenced the erection of those engines to which he has since laid claim as exclusive inventor.*

Shortly after the first experiments were made by the Marquis de Jouffroy, a gentleman of the name of Miller, who resided at Dalswinton, published a work, in which he described the application of wheels to the working of triple vessels on canals; and in 1794 he completed a model of a boat on this construction, impelled by a steam engine.

From this period till 1801, but little progress appears to have been made in this species of navigation: in that year Mr. Symington, who had

---

* The Quarterly Review, in an admirable article inserted in the thirty-eighth number of that work, very justly exposes the pretensions of the Americans to this invention; and points out some of the advantages which society owe to the above *modest* and *philanthropic* individual, not the least of which is the attempted introduction of the *torpedo,* and other apparatus for destroying human life by wholesale.

been employed in the construction of Miller's vessel, tried a boat propelled by steam on the Forth-and-Clyde Inland Navigation: this, however, was shortly laid aside, on account of the injury with which it threatened the banks of the canal, from the violent agitation produced by the paddle-wheels.

Mr. Symington's mode of connecting the piston and paddle-wheel, was by placing the cylinder nearly in an horizontal position, so that by this means the necessity of employing a working beam was avoided. The piston was also supported in its position by friction-wheels, and communicated, by means of a rod, with a crank connected with the wheel, which imparts a motion to a paddle somewhat slower than its own. The paddle-wheel was placed in the middle of the boat towards the stern, and on this account it became necessary to have a double rudder, connected by rods, which were moved by a winch placed at the head of the boat.

Mr. Symington also employed stampers placed at the head of the boat, for the purpose of breaking the ice on canals; and this plan, we believe, was also adopted in the original construction of the vessels intended for the Arctic expedition.

In 1795, a very ingenious apparatus was invented by Lord Stanhope, and tried by that nobleman in Greenland Dock. In this experiment, the paddles were made to resemble the feet of a duck, and were placed under the quarters of the vessel. This plan was also tried in America, but it does not appear in either case to have answered the expectations of its projector.

A plan has also been tried which in some measure resembles the French chain-pump. This was, we believe, first employed in the Duke of Bridgewater's canal, and consists in the use of a chain, with a number of paddles attached to it, going over two wheels placed level with the water line. A steam engine acting on the foremost roller, gave motion to the chain, and a continuous parallel motion was thus effected.

In 1800 Mr. Linnaker obtained a patent for propelling vessels by forcing a stream of water from the stern, an additional supply being at the same time drawn in at the head of the vessel. This ingenious contrivance, however, appears to have been practised nearly a century back: a very circumstantial account of the apparatus for this purpose being prefixed to the *Specimina Ichnographica* of John Allen, published in 1730.

The first really practicable, and we may add profitable attempt at steam navigation in Europe, appears to have been made on the Clyde in the year 1812. This was a vessel for the conveyance of passengers, with an engine of only three horses' power, and which was of considerable draught. On account, however, of the numerous shallows in our rivers, it has since been found advisable to construct the vessels employed in this species of navigation so as to draw as little water as possible.

The greatest number of boats now in use either on the Thames or Clyde navigation, are fitted up for the conveyance of passengers. They have two cabins, one before the engine, which is smaller, and at a reduced price, while the second, or large cabin, is usually fitted up in the most elegant manner. In some cases the cabins enjoy the additional advantage of being heated by steam, while others are heated by means of a pump, which forces a current of air over the chimney into the cabin. The engine-room is seldom more than twenty feet in length, and little more than half as many in width; this being sufficiently large for an engine of twenty horses' power, with all requisite apparatus, two boilers, and abundant stowage for coals. By an ingenious contrivance, the chimney, which is of considerable height, is made to lower nearly level with the deck. The joint that covers the flue during this process, acting upon the same principle as the sliding shells of a lobster's back, which completely prevents the escape of smoke.

From a series of accurate experiments and calculations lately made, it appears, that the expense attendant on the navigation of a small vessel is much larger in proportion, than where an engine of greater power is employed; and, consequently, we find that steam-boats of great burden are now constructed. In America, more particularly, these boats usually run from three to four hundred tons burden, the great width of their rivers rendering vessels of this size perfectly manageable. Mr. Buchanan, however, recommends the use of a vessel, whose dimensions shall not exceed seventy feet in keel, and ninety tons burden, as the most eligible size for a luggage and passage boat.

It has been stated, and we think with some degree of reason, that the security and comfort of passengers by steam-boats would be best consulted by the employment of a subsidiary vessel of sufficient burden to convey the engine and heavy luggage. By the adoption of this plan, the danger to the passengers from the bursting of the boiler, or other apparatus, would be entirely prevented, and the high-pressure engine generally employed. It would also remedy the loss and inconvenience attendant on the continual shaking of the vessel, which, if large and lightly constructed, will be shaken asunder by the engine in a very short period. And last, though certainly not the least among the advantages likely to accrue from the adoption of this plan, it would materially diminish the disagreeable effect arising from the heat of the furnace and noise of the engine.

To remedy the inconvenience that has sometimes been experienced from the difference in the draught of vessels by which the paddle-wheels are at one time sunk beneath the centre of their axis, while at another time they scarcely touch the water, a mode of adjusting their relative height has been introduced. An apparatus calculated to effect this very desirable object, without retarding the motion of the vessel, was lately presented to the Society of Arts, by Mr. Dickson, who received the silver medal for the above communication.

Mr. D. states, that the great utility of this improvement consists in its application to sailing vessels: for instance, suppose a steam vessel to be going direct against the wind by means of the whole power of her steam engine, and that the wind should change and become favourable, the propellers may, by these means, be immediately raised out of the water, and the vessel allowed to have the whole effect of the sails, thereby saving the expense of fuel. All steam vessels now in use, experience so great an impediment from the propellers being always in the water, as to render sails of no benefit. Another advantage will be derived when there is only a gentle breeze in the vessel's favour, as the propellers can be set to work, which will take hold of the water at pleasure, and thereby unite the power of the steam to that of the wind, which will secure the passage in the given time, at much less expense, as the engine will only consume fuel in proportion to the labour it has to perform. A farther advantage will be found when the vessel has only a side wind; for, by the use of this contrivance, one of the propelling wheels can be worked with its full power in the water, and the other entirely lifted out, if necessary.

For the purpose of accurately trimming the vessel, Mr. Dodd recommends the use of a tank or cistern, placed beneath the projecting deck on each side the vessel, the water ballast for which may be raised by the working of the engine. This, on being discharged by means of a plug, would give a preponderance to the opposite side of the vessel, and as it would be placed at the end of a lever, the fulcrum of which is the keel, a small quantity of water would have considerable effect. A contrivance nearly similar to this, has been adopted for a considerable period of time on board one of the Gravesend steam vessels. It consists in the use of a small carriage made to contain an iron cable, which is occasionally employed to moor the vessel, and its weight is such that the vessel may readily be trimmed by moving it from side to side.

Mr. Maudslay has lately constructed a large engine for a steam-boat invented by Mr. Brunel, which has two cylinders acting alternately upon different cranks, formed upon the same axis at right angles to each other, so that the motion is continued without the action of a fly-wheel. In this engine, one boiler is placed between the two cylinders, and one air-pump and condenser exhaust them both; so that by these means an engine of considerable power is contained in the smallest possible space.

Some idea of the prevalence of steam navigation in the more northern parts of our island, may be formed from the following estimate of the number of passengers who have availed themselves of this species of conveyance in the course of one year. On the Forth and Clyde canal, between Glasgow and Edinburgh, 94,250; between Glasgow and Paisley, by the Ardrassan Canal, 51,700; and from Glasgow, along the Monkland canal, 18,000.

Steam-boats of a large size are now employed in the Adriatic. One (La Carolina) goes regularly every second day from Venice to Trieste; another (L' Eridano) passes between Pavia and Venice, and with such celerity, that the voyage is accomplished in thirty-seven hours.

We have now to notice the labours of our transatlantic brethren in this important branch of naval engineering. Profiting by the hints thrown out both by the Marquis de Jouffroy and Mr. Miller, Fulton, who had also seen Symington's boat, ordered an engine capable of propelling a vessel to be constructed by Messrs. Boulton and Watt. This was sent out to America and embarked on the Hudson in 1807,* and such was the ardour of the

* Its first appearance on these waters, is thus described by the biographer of Fulton.

"She had the most terrific appearance, from other vessels which were navigating the river, when she was making her passage. The first steam-boats, as others yet do, used dry pine wood for fuel, which sends forth a column of ignited vapour many feet above the flue, and whenever the fire is stirred a galaxy of sparks fly off, and in the night have a very brilliant and beautiful appearance. This uncommon light first attracted the attention of the crews of other vessels. Notwithstanding the wind and tide were adverse to its approach they saw with astonish-

Americans in support of this apparently new discovery, that the immense rivers of the new world, whose great width gave them considerable advantages over the canals and narrower streams of Europe, were soon navigated by these vessels.

The city of New York alone possesses seven steam-boats, for commerce and passengers. One of those on the Mississippi passes two thousand miles in twenty-one days, and this too against the current which is perpetually running down. The above boat is 126 feet in length, and carries 460 tons, at a very shallow draft of water, and conveys from New Orleans, whole ships' cargoes into the interior of the country, as well as passengers.

The following list of steam-boats now in operation on the river Mississippi, and its tributary streams, has been published by Mr. Robinson.

	Tons.		Tons.
Vesuvius	390	Kentucky	80
Etna	390	Governor Shelby	120
Buffalo	300	Madison	200
James Monroe	90	Ohio	443
Washington	400	Napoleon	332
Constitution	75	Volcano	250
Harnot	40	General Jackson	200

ment that it was rapidly coming towards them; and when it came so near as that the noise of the machinery and paddles was heard, the crews in some instances shrunk beneath their decks from the terrific sight, and left their vessels to go on shore, while others prostrated themselves, and besought Providence to protect them from the approaches of the horrible monster, which was marching on the tides, and lighting its path by the fires which it vomited.

	Tons.		Tons.
Eagle	70	Experiment	40
Hecla	70	St. Louis	220
Henderson	85	Vesta	100
Johnston	80	Rifleman	250
Cincinnati	120	Alabama	200
Exchange	200	Rising States	150
Louisiana	54	General Pike	250
James Ross	320	Independence	300
Frankfort	320	Paragon	400
Tamerlane	320	Maysville	150
Cedar Branch	250	Total	7259

Building.

	Tons.		Tons.
2 at Pittsburgh of 180 tons	360	1 at Portland (Kentucky)	300
2 at Wheeling, of 500 and 100	600	3 at New Albany each 220	660
2 at Steubenville	90	4 at Clarkesville	500
1 at Marietta	130	1 at Salt River	160
1 at Maysville	110	1 at Vevay	110
2 at Cincinnati	720	1 at Madison	120
2 at Cincinnati 115 and 250	365	1 at Rising Sun	90
2 at Newport	500	1 on the Wabash	80
1 at Jeffersonville	700	2 at New Orleans each 200	400
		Total	5995

In addition to which there have been lost by accidents of different kinds, the following steam-boats: Orleans, 400 tons; Comet, 15; Enterprise, 45; Dispatch, 25; Franklin, 125; Pike, 25; New Orleans, 300.

The Savannah, of 350 tons burden, crossed the Atlantic, and arrived at Liverpool from the United States in twenty days, the greater part of which time the steam engine was in action. The steam apparatus in this vessel occupies the greater part of the hold from the main-mast to the fore-mast, a small space being reserved at each side for the conveyance of coals which in this engine amounts to about ten tons per day, and from this it will be seen that there is but little room can possibly be left for the stowage of cargo, &c.*

The paddle-wheels in the Savannah, are affixed to a cast-iron axle-tree, passing through the sides of the vessel above the bends; nearly the whole of each wheel being taken to pieces, and removed in the event of bad weather: two principal arms, which are also of cast metal, being the only parts remaining, and these in high seas are placed in a horizontal position, producing but little inconvenience in the navigation of the vessel.

According to Mr. Buchanan the expense of a steam-boat on the Clyde navigation, with two engines, each of which make forty-five twenty-two inch strokes per minute, appears to be rather more than two thousand pounds.† The following estimate contains the separate items.

---

* Vide Act regulating the Stowage of Steam-vessels, in Appendix.

† The number of hands employed in this vessel were eight; viz. captain, pilot, engine-man and assistant, seaman and assistant, and the steward and assistant. Those carrying goods only, have seldom more than five hands.

The hull	£700
Steam Engine	700
Paddle wheels and other machinery ...	300
Joiner's work	200
Upholsterers' work	70
	1970
Contingent expenses	330
	2300

Mr. Dodd estimates the expense of completing a vessel of an hundred tons burden, and drawing four feet six inches water, at about six thousand pounds; while the first cost of a steam-boat on Mr. Symington's plan, doing the work of about twelve horses, and travelling at the rate of two miles and a half per hour, has been estimated at about eight or nine hundred pounds. The consumption of coals however, in this and in all other engines connected with steam navigation is much more considerable than when employed on land. In a fourteen-horse power engine, on Messrs. Boulton and Watts construction, they consume 1 cwt. 1 qr. 20 lbs. per hour of good Newcastle coal: while an engine of thirty-three horses' power, requires a proportion scarcely equal to two-thirds that quantity.

The employment of high-pressure engines for the purpose of steam navigation, has unfortunately given rise to a considerable prejudice against the general use of this economical and expeditious prime mover. The attention of the legislature was first drawn to this subject by the explosion of a high-pressure boiler in a vessel employed for the conveyance of passengers in the neighbourhood of Norwich, on which occasion the consequences were of the most terrific nature. As, however, the matter was very fully examined by the Committee of the House of Commons, appointed to inquire into the various particulars connected with this unfortunate catastrophe, we shall content ourselves by referring the reader to the important facts contained in the annexed abstract, which fully demonstrates the possibility of constructing a steam vessel, uniting the essential qualities of safety, economy, and celerity.

CHAPTER IV.

Abstract of Evidence before a Select Committee of the House of Commons on STEAM NAVIGATION.*

Mr. BRYAN DONKIN was called in and examined.

WITNESS went down to Norwich, as a volunteer, to inquire into the cause of the explosion of a steam-boat. Was accompanied by Mr. Timothy Bramah and Mr. Collinge. Was of opinion that the immediate cause of that explosion had been the use

* The Committee commenced its sittings May 8, 1817, and consisted of the following highly respectable individuals:— Charles Harvey, Esq. in the chair; Mr. William Smith, Mr. Davies Gilbert, Sir Martin Folkes, Sir James Shaw, Sir William Curtis, Sir Charles Pole, Mr. Alderman Atkins, Mr. Williams Wynn, Sir Edward Kerrison, Mr. Lacon, Mr. Shaw Lefevre, General Thornton, Mr. Edward Littleton, Mr. Finlay, Mr. Leader, Mr. Alderman Smith, Mr. Wrottesley, Mr. Barclay, Sir James Graham, Mr. Swann, Mr. Charles Dundas, Mr. Holmes, Mr. Thompson, and Mr Bennet.

On the 14th of the same month, Sir Matthew Ridley and Mr. Ellison were added to the Committee.

of steam of a very high expansive force; the approximate cause was a deficiency in strength of the end of the boiler. The boiler was cylindrical. The cylindrical part, and one end, was wrought iron; and the other end was cast iron. It appeared to have been previously of wrought iron, but, for some reason, the wrought iron end had been cut out, and a cast iron end substituted in its place.—Was of opinion that any high-pressure boiler so constructed was unsafe. The difficulty of obtaining a proper degree of strength at all times, in the materials of which the boilers were made, arose from the constant deterioration which they must be suffering from the action of the fire, and from the various degrees of expansion and contraction, operating on different parts of the boiler.—Would not choose to use a high-pressure engine, from the danger which arose from their use.—Thought it just to state to the Committee, that there was an advantage to be derived from the use of high-pressure engines on board of boats, which were necessarily loaded differently at different times. This different loading required a different power in the steam engine, and the high-pressure engine was capable of having the additional power given to it without difficulty; whereas, in the lower-pressure engines, they were confined to the power first assigned them.—Scarcely ever saw the low-pressure engine beyond six pounds to the inch.—Had known one boiler worn out in six months, and another used for seven or fourteen years. The strength of cast-iron boilers was extremely uncertain: cast iron was liable to contract in various degrees in different places, and therefore was liable to break.—Thought that all cast-iron boilers were dangerous when used for steam of high expansive force. It was more practicable to make a boiler of the malleable metals to resist a high pressure, as far as the tenacity of the metals was concerned; but another difficulty occurred which prevented the application of the malleable metals to boilers for high-pressure engines, which was that of rendering the joining of the plates secure.—Believed that wrought-iron boilers were much less frequent than the cast-iron boilers, and in Wolfe's engines they were scarcely used at all.—Should think that the cast-iron boilers would be cheaper than wrought, if made of equal strength.—Considered that in case of the explosion of a cast-iron or a wrought-iron boiler, the cast-iron would be attended with the greatest danger. In employing the malleable metals a simple rending generally took place, so that it would seldom happen that the upper part of the boiler would be torn off; but, in a cast-iron boiler, the fragments would be scattered about, and be more destructive.—The boilers invented by Mr. Linns and Mr. Wolfe were all of them cast iron. —Mr. Wolfe's had been in use nearly ten years. Considered low-pressure boilers as safe from explosion in all instances, used with no farther pressure than six pounds.—Had seen very few boilers constructed for the purpose of a low-pressure engine, or a condensing engine, that would sustain a pressure of ten pounds without occasioning considerable leakage, or without forcing the joints.—Had never heard of an explosion with the low-pressure boiler of any consequence whatever,—merely giving way of the plates, or the wearing out, not such a bursting as could be called an explosion.—Conceived Wolfe's mode of constructing boilers to be a considerable improvement,—a very material one. Had likewise been told, though without having seen one, that Trevethick had invented a method of making boilers by increasing their length and decreasing their diameter, so as to render them capable of sustaining pressure to a much greater degree than heretofore.—In high-pressure engines the expansive force of steam was very variable, from thirty pounds to one hundred and twenty pounds upon the square inch, or even perhaps higher than that. Instances had been known in which a boiler had been worked at one hundred and sixty and one hundred and eighty pounds.—Had no doubt but Cornwall had derived incalculable advantages from the use of high-pressure engines.—According to the general construction of low-pressure boilers, they were so rivetted together, as to withstand the low pressure they were intended to bear; and they always gave indications of an increase of pressure long before any danger could be apprehended from them either by the joints giving way, or the steam forcing a passage through.—Had witnessed several experiments on Wolfe's engines, where the object was to ascertain the comparative expenditure of coals or fuel in grinding corn, between his engines and the low-pressure or condensing engines; and the results were decidedly in favour of Mr. Wolfe's engines.—Apprehended that there was no saving of fuel, or very little, in the common high-pressure engine.—The average effect in Wolfe's engine, was the grinding eighteen bushels of wheat with one bushel of coals; while the average effect of Boulton and Watt's engine, or the low-pressure engines, was the grinding of from ten to twelve bushels of wheat with a bushel of coals.

SETH-HUNT, Esq. was called in, and examined.

Had formerly been commandant of Upper Louisiana.—Knew that in the United States a great number of steam-boats had been established. The first was at New York. There were then running between New York and Albany, ten boats: two between New York and the state of Connecticut; four or five to New Jersey; besides the ferry-boats, of which there were four. These boats were all worked by low-pressure engines: no accident had ever happened to any one of them: they had been running since the year 1807; and the boats at Albany performed about forty trips each per annum.—They went a distance of a hundred and sixty miles in twenty-one hours, and came down in nineteen: sometimes a little longer, but never shorter than nineteen; that was the quickest passage.—Some of them went about seven miles an hour in still water: some boats had gone nine, ten, or eleven miles; but that was under particular circumstances. They had come from Newhaven to New York (ninety miles) in six hours and a half, without any sail.—Those which went to Albany passed up the North River; and the others, to Connecticut, passed through Long Island Sound, forty miles broad in one part of it. On the river Delaware there were a number of boats also established, which plied between Philadelphia and Trenton in New Jersey; also others between Philadelphia and Newcastle, and Philadelphia and Wilmington, beside ferry-boats. Several of those boats had low-pressure engines; others had high-pressure engines, from one hundred to one hundred and forty pounds on the square inch, and as high as one hundred and sixty; but those engines were constructed upon Oliver Evans's plan, called the Columbian plan. They were of wrought iron.—There were no boilers cast in America. Presumed that might arise from their not having founderies in which they could cast them sufficiently large. They were all wrought-iron boilers, or copper: all which had to pass through salt water were copper. The boat Etna, which passed between Philadelphia and Wilmington, was a high-pressure engine, and outstripped all the other boats: there was no competition at all between them. There were boats from Baltimore to Norfolk, which passed a part of the Chesapeak, sixty miles in width. They have been to New London, which was still more exposed; and had been up to New Hertford. Those were low-pressure engines.—The Powhawton steam-boat was built at New York; went into the open ocean; encountered for three days a very severe gale of wind, arrived safe at Norfolk, and up to Richmond. The gentleman was now in England who navigated her; and had heard him say that he felt himself as safe as he should in a frigate; and he said there was this advantage, that the steam power enabled him, when they could not have borne sails, to put the head of the vessel to the sea instead of lying in the trough of the sea, being exposed to be overrun by the waves.—The largest steam-boats in America were those on the Mississippi, the Etna and the Vesuvius which ply between New Orleans and Naches. They were four hundred and fifty tons, and they carried two hundred and eighty tons merchandise, one hundred passengers, and seven hundred bales of cotton, besides the passengers transported to New Orleans.—Remembered only three accidents having happened to steam-boats in America. The first happened on the Ohio, and was occasioned by the negligence and inattention of the engineer, who loaded the safety valve, and neglected to attend the fire. All hands were engaged in hoisting the anchor: the fire was in a very high state, and of course produced a vast deal of steam that did not escape by the ordinary operation of the engine, which would discharge it and carry it off.—What was called the safety-valve had been improperly loaded and neglected.—The next accident happened not from any fault of any body, but from an act of God: it was lightning, as was satisfactorily explained to the public, both by the passengers, and those interested in the boat. That was at Charlestown, in South Carolina. The pipe which carried the smoke up to the top attracted the lightning, and it went down, and split the boiler.—A third accident happened lately to the Powhawton. She was not in operation when it happened: they were out of fuel; they stopped their boat, and lay still upon the water, while they went after wood; still, however, they kept up their fire; and the steam was so high, that it exploded in that situation, there being no consumption of the steam as it accumulated. Those are the only accidents that ever happened, except such as have happened from vessels taking fire.—No accident had ever happened in America to a high-pressure engine, either in a manufactory or out of it; and there were many engines used in the manufactories, and in flour-mills and saw-mills, constructed upon the plan of Oliver Evans, which acted on the high-pressure principle to one hundred and fifty pounds an inch. He had worked one hundred and sixty, but one hundred and twenty was his constant average. The fuel, in most places, was wood; at Pittsburgh, and on the Ohio river, it was coal and wood; at Pittsburgh, and at Laceling, and at a hundred other places, there was a solid mass of coal fifty miles square. They drove a shaft horizontally into the hill, and the coals were abundant above their head; in the mountains, as fine coal as any in the world. It was delivered at the houses of the inhabitants at sixteen bushels per dollar.—The number of steam-boats was rapidly increasing upon both the low and high pressure system, because they had different interests and different companies. Mr. Evans being a patentee, they had to give something for the use of his patent; if they could not make their bargain with him, they used the low-pressure engine; but there was a new engine, built for one-third of the money, coming into use in several of the steam-boats, invented in America, a perfect rotatory engine; and it was supposed that it would supersede all other engines.—Knew of no other guard than that of properly constructing the safety valves, and the manner of loading them, so that they could not get on more than a certain weight; they must of course construct them strong enough, and prove them. They were under no Government regulations.—It was supposed that a rotatory engine consumed less coals than one with a reciprocating beam. Twelve bushels of coals, with the rotatory motion, would perform the same work as the other engine with twenty.

Mr. TIMOTHY BRAMAH, of Pimlico, Engineer, called in and examined.

Did not think that a high-pressure engine, under any guard that could be applied to it, was a safe engine to use in a steam-boat.—Thought that if a boiler was prepared to sustain one hundred pounds, and strained with a force equal to two hundred pounds, it might afterwards, perhaps, break at forty, the straining having injured it.—Apprehended that a boiler, upon a proper construction, of wrought metal, might be tried with a certain force, so small in comparison with that pressure which it was intended to bear, as not to incur any risk of being injured in the proof, and have a complete surplus of strength, so as to enable it to be afterwards used without any danger in the use.—Would recommend the use of two safety-valves, one to lock up; and to have it examined once a week, or as often as might be necessary, to see that its action was perfect.—Had seen many cast-iron vessels burst. The wrought iron generally tore and opened out, to admit of the fluid escaping; it was generally the fluid which did the mischief when the wrought iron was used, and it was both the fluid and the materials which did the mischief when the cast iron burst. The effect in cast metal was to carry the pieces of the metal to a considerable distance, which was seldom the case in the wrought, unless where there was any cold shut in the metal. The cast would burst like a shell, projecting the particles of the metal to a considerable distance.—Where wrought-iron boilers had burst, the injury sustained by the individuals had arisen from the fluid's escaping, and in cast-iron boilers it had frequently been by the expansion of metal. Copper being purer was not subject to the same danger.

Mr. JOHN TAYLOR, of Stratford, in Essex, called in and examined.

Was acquainted with the accident which lately happened to the steam-boat at Norwich. Had heard that the plate of cast iron was of inadequate thickness for the strain to be put upon it. With respect to the impropriety of cast iron compared with wrought, had also constructed one of the high-pressure boilers precisely in the same manner: the boiler was proved to one hundred pounds a square inch by the water-proof, commonly used with about forty pounds' pressure; but the cast iron broke one day with less than twenty pounds' pressure of steam; the fracture being caused evidently by the heat expanding unequally, and being kept from going to the form it would otherwise assume.—Had seen the Well-street boiler intended to boil sugar. The thickness was intended to be about two inches, or two inches and a quarter; but by inserting the core unequally, the thickness on one side was three quarters of an inch; on the other side the thickness of the metal was two inches and a quarter, or thereabouts; therefore, to the general difficulty of cast iron was added a most improper construction.—Understood from the men who were working there, that there had been something like a mercury gauge attached to it, but that the mercury never fluctuated. It was probable there was a pressure of more than one hundred pounds. — Considered that a wrought-iron boiler might be rendered safe by the use of a column of mercury in a siphon or tube of sufficient size. When that mercury was displaced by the expansive force of the steam, which would be regulated by the height of that tube, to admit of the efflux of the steam from the boiler as fast as it was generated by the fire; in that case, the expansive force could not increase in the boiler, the mercury would be blown out, and the steam would escape. Conceived it essential to have a second safety-valve, which should be under the control of the master or proprietor of the works; and there was another small contrivance very important to the safety of the boiler. Boilers had frequently been weakened very much by the water having been evaporated too low. To remedy this, a hole should be previously bored in the bottom, rivetted by a piece of lead, so that the lead would remain perfectly secure as long as it was covered with water; but, the moment the water left it, the lead would melt, and the steam being blown through the hole, would put out the fire. Besides giving the signal of what was wanted, it would at once put an end to the cause of danger.—Considered that the mercurial gauge acted as a safety-valve, which could not be stopped or put out of order; and it had the advantage of exhibiting, during all times of the boiler's working, the state of the steam within the boiler, by the fluctuation that took place in that column, as indicated by the appearance upon the surface of the mercury. If the mercury became stationary, it might readily be suspected that that tube was stopped; therefore it would point out itself instantly that it had become not what it ought to be. The safety-valve had not that advantage, as it did not indicate any thing till the steam was blown out by raising the weight. —With respect to the value of high-pressure steam for working engines in Cornwall, of late a a most valuable improvement had taken place; and if it was an object to save coal to steam-vessels upon a large scale, high-pressure steam became an object of great importance to them, if applied upon the principle that Mr. Wolfe had in the first place introduced, but which had been applied by Mr. Simms and others. Was of opinion that those high-pressure boilers might be made with equal safety as low-pressure boilers.—Had prepared a statement of the work done by the engines on the principal mines in the county of Cornwall. It stated the consumption of coal and the work done by every engine therein named; from which it appeared that the average work of engines then in the county of Cornwall was to raise about twenty millions of pounds of water one foot high by the consumption of one bushel of coals: that, by the introduction of high-pressure steam, under the best mode of management, an effect equal to from forty-three to forty-five millions pounds of water was raised to the same height by the same quantity of coal, thereby producing above double the effect.—Apprehended that condensing or low-pressure engines were equally liable to be blown up by the carelessness and inattention of the engineer conducting them with high-pressure engines. In France, at Crusal, some very good engines were erected by Mr. Wilkinson, at a very large work; they were on Boulton and Watt's principle: one of them blew up, and killed several people.—Conceived that the mercurial gauge, if of sufficient bore, might be applied with ease to the high-pressure boilers, so as to produce safety as certainly as the column of water, which was in fact a water-gauge, such as was usually applied to the low pressure.—Conceived that there would be no difficulty in constructing a safety-valve, so as to operate with certainty, and yet be safe from any impediment which the engineer might intentionally place in the way of its operation, without incurring any very considerable expense.

Mr. John Collinge, of Bridge Road, Lambeth, Engineer, examined.

Went to Norwich, in consequence of the accident that happened to the steam-boat there.—Attributed the explosion of that engine to the construction of the boiler. It was composed entirely of wrought iron, except one end, and that was capped with cast iron. The cylindrical part was made of wrought iron.—Was of opinion that any material under very severe pressure was liable to fail, and cast iron for this reason, because, in all large bodies, it was found that the air could not wholly escape in the act of fusion. Had occasionally had large masses of cylinders and pans to break up, and frequently found cells where the air could not escape. There was certainly a much greater dependence upon wrought iron or upon wrought metal; perhaps it would be better to include copper.—In the event of any accident happening to the boiler, the greatest mischief would be likely to arise from cast iron, because cast iron flew off in fragments, and wrought iron from its superior tenacity did not.—Remembered an accident having occurred at Malden, where a boiler, nineteen feet long, was blown off from the seat of its connection with the base. Had found, in making wrought-iron boilers, that, if they were made of metal of a considerable substance, they could not be so well united to make them steam-tight; it was a very difficult thing to do. The rivets that were applied to wrought-iron boilers were put in hot, and, when they were hammered, to secure the joint, they got cold, shrunk, and did not always fill the hole through which they had passed.—Had no conception that any safety-valve could be applied to render them perfectly secure under heavy pressure.—Thought the mercurial gauge would be the greatest safety for a boat, if it could be judiciously applied.—The condensing engines should not be more than four pounds to an inch; and, if the capacity of the vessel allowed of it, the condensing engines answered every purpose, because the making a wrought-iron boiler would be on such a scale of thickness, that, if more than the usual pressure was applied, the rivets would fail, and constitute a security against any fatal occurrence.—Thought from the power that was wanted into steam-boats, condensing engines were the best engines applicable for that purpose.—Did not conceive it impossible to construct a wrought-metal boiler, with safety-valves properly adjusted to its capacity, and a mercurial gauge, supposing that to be capable of being applied, which should render a high-pressure engine, on board a steam-boat, what might be called perfectly safe.—Thought that in order to give security to the public in travelling by steam-boats, it might be necessary to have an examination of each engine two or three times a year, as it would create confidence.

Mr. William Chapman, of Newcastle-upon-Tyne, Civil Engineer, examined.

Considered all engines, whether high-pressure or low-pressure, as dangerous to the passengers, unless due precaution was taken to emit the steam, when exceeding a given pressure; for, in low-pressure engines, the boilers were always liable to burst, or to alter their force, when the pressure became greater than the resistance. All wrought-iron boilers, but those that were cylindrical in the section, and with hemispherical ends, or portions of spheres, or cones, or coniads, were liable to alter the form by the natural expansive force of the steam, and therefore all boilers of those forms owed their safety to their weakness, because, if weak, they would alter their form without danger, and, if strong, they have been known to bend the iron so abruptly as to break asunder. — There were high-pressure engines, working with a force of from fifty to sixty-five pounds per inch, and no accident had happened to any of them but to one, the safety-valve of which was stopped up by a man sitting upon it purposely. He said, he would have a good start, and surprise them. The consequence was, the boiler blew up, and killed and wounded a very considerable number of people.—Considered that the high-pressure engine could only be rendered safe by having the boiler of the form already described, and the cylindrical part of an unlimited diameter, with a competent thickness of wrought iron or copper, and the plates secured to each other by a double line of rivets. It was also requisite that there should be two safety-valves, each laden with any determinate weight per superficial inch of the narrowest part of the seat of the valve. One of those valves should be at perfect liberty to be raised at the pleasure of the manager, because sometimes it was expedient to raise it. The other should be under a cover of such description as not to be opened at all at the discretion of the engineer, but with sufficient apertures for the emission of the steam, and for any of the passengers to see that the valve was not made fast. It was also requisite that there should be a mercurial gauge of not less than an inch in diameter, and whose longest limb should not be greater than two inches and one-eighth for every pound per inch upon the safety-valve for each. It was necessary, by occasional inspection, to take care that the mercury did not stiffen by oxydation, occasioned by the heat and motion to which it was in a slight degree liable.—Conceived that a high-pressure engine, thus guarded, might be used with perfect safety on board a steam-boat, so long as the boiler was kept in order; but the boiler bottom was liable to erode or consume by the action of the fire, and therefore required watching.—Thought a boiler might last twelve months, provided its bottom was made of charcoal-iron, beat, not rolled, because there was a great deal of difference in the grain.—Would recommend all the boilers on board steam-boats to be made either of copper, or charcoal-iron plates, beat under the hammer, and not rolled. The resistance of the cylindrical boilers would be precisely in the inverse ratio of the diameter.

Mr. PHILIP TAYLOR, of Bromley, Middlesex, Manufacturing Chemist, examined.

Considered the first and most material point to attend to in the construction of high-pressure boilers was, that their diameters should be small in proportion to their capacity; that as small a proportion of the external surface of the boiler as possible should be exposed to the destructive action of the fire; and that that portion of the boiler which was subjected to the action of the fire should be so situate and guarded, that, in case of explosion, the least possible mischief should arise. In all boilers which he had made use of, no portion of the boiler was exposed to the action of the fire, without its being constantly covered with water. In the boilers constructed under his direction, the fire was applied under an arch of not more than two feet and a half in diameter: this provided against any extensive rents taking place in the event of explosion. All the boilers he had hitherto employed had been constructed of malleable iron, commonly known by the name of charcoal-iron, rivetted together, and secured by strong wrought-iron bolts. From observing the danger arising from the introduction of flat cast iron ends, had invariably terminated the ends of the boilers by wrought-iron ones, nearly hemispherical: this mode of construction, so far as his experience had gone, combined more strength and durability than any other. The precautions he had used to guard against the nuisance of such boilers, had been by adapting to them two safety-valves; one under the control of the engine-man, the other secured in a strong cast-iron case, locked down, and loaded with such a weight, as would suffer the steam to escape when it had arrived at an improper degree of expansive force; safety-valves not having at all times answered the purpose intended. Had, likewise, in every instance, attached to the boiler a mercurial column, the bore of which was proportioned to the size of the boiler; and considered an iron tube, of an inch diameter, sufficient to guard against accident, when applied to a boiler four feet in diameter and twenty feet in height; because the limit given by such a column came far within the limit of absolute safety. The external limb of the mercurial gauge had, in all cases, been proportioned to the strength of the boiler applied, taking care that the expansive force of the steam would displace the mercury long before any dangerous expansive force would arise. In order to guard against the boiler being injured by the action of the fire, from a deficient quantity of water in the boiler, had inserted a leaden rivet in such a situation, that it would melt as soon as it was uncovered by the water, and produce an opening, which would suffer the escape of the steam.—Considered cast-iron boilers safe, provided their various parts were made of small diameters in proportion to their capacity; such, for instance, as those constructed by Mr. Wolfe.—Thought that a boiler constructed on this principle was equally safe with those called condensing engines, because a greater attention to strength is always paid in the construction of high-pressure boilers than in the construction of those for low-pressure engines, in proportion to the pressure they have to sustain.—The high-pressure engine, constructed by Mr. Wolfe, employed not only the expansive force of the steam, but also that power which was acquired by its condensing; and the effect in Cornwall had been, that engines on this construction had done double the quantity of work with the same quantity of fuel.—Should consider any measure tending to impede the use of high-pressure engines injurious to the country.

Mr. Henry Maudslay, of Lambeth, Engineer, examined.

Never considered high-pressure engines were applicable to boats, because the purpose of high-pressure engines was to save water, and water could not be wanted on board a vessel; the difference between the one and the other made no saving, either in the weight or expense, taking it ultimately, particularly when steam-boats were properly contrived.—Built the Regent steam-boat with a low-pressure engine. There was a dispute between two men, and one of them swore that he would blow his boiler up but he would beat the Regent in coming up. The man certainly did exert himself as much as he could, and kept his steam as high as he could get it, and it flew out of the safety-valve very frequently, and he hurt his boiler materially from doing so; but he did not beat the Regent: but, if it had been a high-pressure engine, he would either have beat her, or blown up his boiler, because he had the power in his own hand. Had employed two and sometimes three safety-valves: to make it quite easy for the man to move them, had a sort of bell pull to pull it up every hour, if he pleased, to keep it in action; because, it was clear, the spindle might corrode and stick fast for want of use. Supposing it not touched once a week, it would not be a safety-valve any longer, because a very little friction would add a great many pounds weight to the opposition the steam ought to meet with.—Never knew a low-pressure engine unsafe, but it appeared that high-pressure engines had been.—Conceived that the same motive which would induce the engineer to work it with an improper pressure, would induce him to leave it untouched, that it might have an improper pressure.—Considered that wrought iron was extremely safe, compared to cast iron.

Mr. ALEXANDER GALLAWAY, of Holborn, Engineer, called in and examined.

Would recommend that, for steam-boats, the condensing engines should be used in preference to high-pressure engines; and, for these reasons: In the first place, the great advantage promised from a high-pressure engine was, that it could be worked in a situation where water could not be procured, and therefore it was, for such a situation, a valuable machine; but, in situations where water could be readily procured, it was not so. And in reference to the comparative price between a high-pressure engine and a low-pressure engine, and in reference to the space that it occupied, and in reference to the superintendence that it required, it was decidedly evident no economy was produced. Speaking of it as a matter of safety, it would be necessary to say, that experience had fully proved that the maximum of force to be obtained by a condensing engine was when the steam was rarefied from three to six pounds on the inch. The engine was then by far more efficient than when the steam was rarefied beyond that point. And it would appear equally clear, that whether it was a cast-iron boiler, or a wrought-iron boiler, or a copper boiler, the force of the engine was better performed by steam at three pounds and a half than at any increased expansive force; the boiler being subject only to three instead of six pounds, it must be less liable to explode or burst at that than at an increased expansive force. Would farther say, that every man that was called to work a condensing steam-engine knew, that, when his steam was at three pounds and a half, it performed a greater quantity of labour than at any other time; for, if it was increased, a vast labour was thrown on the air-pump and the condenser, and the engine retarded; therefore, a man had no inducement to increase the expansive force of the steam, knowing that no useful end could be obtained by so doing, but giving himself additional labour, and consuming more fuel, and performing less work. All boilers on board steam-boats should have the fire in the interior of the boiler, because it was of very little importance, when upon the subject of safety, whether the passengers were to be endangered by an explosion, or whether the vessel was to be weakened in its timbers, or essential securities, by the improper application of the fire to the boiler; would therefore recommend that the fire should be contained in the interior of the boiler, and that there should be an additional safety-valve, which should be solely subject to the superintendence of the proprietor, and that the manager of the machine should have no possible access to it. — Would certainly recommend a wrought-metal boiler in preference to a cast-iron boiler; and the reason was clear, that the operation of casting, however skilfully managed, was always an uncertain process.—Thought that if an additional safety-valve was applied to a boiler, and that safety-valve placed beyond the power of being interfered with by any person but the proprietor, then the boiler would be secure from explosion, if the safety-valve should be judiciously loaded; but if that safety-valve was even placed beyond the reach of the operator, and at the same time injudiciously loaded, a calamity might take place, the same as if no such security existed.—Under all the circumstances of the case, would most decidedly recommend a condensing engine; a condensing engine, with a wrought-iron boiler; because, when cast iron became subject to high expansion and contraction, the constant repetition of those effects in a very great degree impaired the strength of the boiler.—Would venture to say, that all engines in steam-boats should be subject to regulation and inspection by competent persons. A steam-boat must have a register; and, before such register should be granted, the engine should be inspected, to see whether it was of a character to deserve its being considered safe.—Was quite satisfied, that, taking for granted that condensing and high-pressure engines were judiciously formed, the one would take as much fuel as the other, there would be no material saving, if any; but if two principles were associated together, as in the case of

Wolfe's engine, there would be a considerable saving.

Mr. John Braithwaite, of the New Road, Fitzroy Square, Engineer, called in and examined.

With respect to high-pressure steam, would engage to make a boiler, or direct one to be made, which would defy any engineer, or other person, to blow up, or burst; and had lately erected five boilers, which he was ready to prove to any gentleman, and even to any engineer, that they could not destroy them.—Recommended to Mr. Martineau, for whom he erected them, that, as there had been an accident in his neighbourhood, he ought to have a boiler to bear three times the pressure he meant to put upon it; and, if it did bear that pressure, and they applied two safety-valves, with a mercurial steam-gauge properly weighted and adjusted (one of those safety-valves being at the will of the person about the boiler, and the other locked up) it would be impossible to explode a boiler of that description. Saw the boiler after it was exploded at Wellclose-square; and also conversed with one of the men that was saved, who said, that he had carried an additional weight to put on the safety-valve just before it exploded; that the mercurial gauge there was plugged up, so that it was useless; besides which, instead of the safety-valve being weighted equal to forty-five pounds, they added a double weight, which increased it to ninety pounds weight upon an inch, and the boiler was very improperly made.—Would recommend wrought-iron boilers in preference to cast on board of steam-boats.

Mr. JOHN HALL, of Dartford, Engineer, called in and examined.

Had only to observe, that he made his boilers of cast iron, and proved them by an hydraulic press, made for the purpose; and had gone as high as two hundred and fifty pounds to an inch, which he considered enough. Nothing happened; and he meant the next time to try what they would bear; and had no doubt they would bear from seven hundred to one thousand pounds to an inch; for he believed they could be made stronger than wrought-iron boilers; wrought-iron boilers being rivetted together, could not be so strong as those cast in a solid mass.—Had a boiler made composed of three tubes on a large one, and two smaller ones below; the lower tubes which were exposed most to the fire, had cracked, generally by cooling, after the engine had done working. Had known that in three or four instances; perhaps in an hour after the engine had done working, the tubes below had cracked, and the other not.—Supposed that in the event of explosion the greatest danger would be from the wrought-iron boiler.—Considered it quite practicable to adjust a safety-valve to a boiler, which should not be accessible to the engineer, but which should sufficiently protect the boiler from mischief, and which once adjusted, would always act, and might always be depended upon.

Mr. ALEXANDER TILLOCH, of Islington, called in and examined.

Was of opinion, that attending to what should be attended to in every steam engine, and employing proper engineers, a steam engine would be perfectly safe, whether with high-pressure or low-pressure steam. The boilers ought always to be furnished with safety-valves, one of which should be covered, and out of reach, with a box over it, but perforated, so that it might be seen when the steam operated on it. A mercurial-valve is also very good, that is, an inverted syphon, with a column of mercury, proportioned to the purposes for which it is to be employed.—Did not apprehend much danger to arise, in case of explosion, from the mercury, because the tube being always perpendicular, the mercury, when shot out, would fall down in rain. Was of opinion a boiler might be made safe, either of wrought or cast iron; but, for great strain, would prefer cast iron, contrary to the opinion of many people; and the reason for this preference was the same for which it was preferred in making cannon. It was not possible to get thick plates of wrought iron perfect throughout, and it was necessary to trust at last to rivets in joining them; but cast-iron boilers could be made of any strength. Instead of having a boiler that would stand sixty, it might be made to stand six hundred of either wrought or cast iron. Another reason why he would prefer cast iron was, that the sheet iron corroded much quicker, and was destroyed by oxydation, so that a boiler might be safe when first set up, and stand its proof, but very soon become unserviceable, or at least, comparatively so. Boilers should always be cylindrical, and for an obvious reason: capacity should be got by length and number, rather than by diameter. There was no more danger to be apprehended from steam, as to bursting, than from the employment of condensed air, only that the water might scald: but, as to the danger of the fragments being scattered about, it was the same with air as with steam, and yet all the engineers constantly employ cast-iron receivers, condensers, or air-vessels, where pressure was wanted.—In case of actual explosion should think the greatest mischief would arise from the cast-iron boiler.—Was aware that there might be cavities in cast iron, but a boiler being proved to a strain beyond that it was to be exposed to by heat, the safety of the boiler was secured, for the temperature never could be at that point which would endanger a fracture from that circumstance.

Mr. George Dodd, Civil Engineer, of Oxford Street,

Stated, that out of five steam-boats under his direction, only two had suffered by partial accidents, and these were owing to the carelessness of the engine workers. His boilers were made flat sided, with flat and dome roofs, the largest of them containing at least fifteen hundred rivets, each of which in some measure answered the purpose of a safety-valve.—Was of opinion that to all boilers there should be two safety-valves. The one which would be accessible to the engine worker should be loaded with the minimum of the pressure that the chief engineer saw fit that the boiler should sustain; and that the one which would be inaccessible and locked up should be loaded equal to the ultimatum that he would, under any circumstances, permit the boiler to support.—Would not allow the safety-valves to be loaded with more than half the weight which had been previously tried, and found the boiler was capable of supporting.—Was of opinion that a boiler whose sides and ends were flat, if properly constructed, and of sufficient thickness in the plates of wrought iron, might be safely used on board steam-boats having the low-pressure engine.—In the Richmond steam-boat the fire was entirely surrounded by the water. It was the case also in the Majestic; but, in the Thames, and in the new boat to Richmond, and the new boat to Gravesend, they were what was called open furnace-mouths. Under the furnace-mouth was placed an ash-hole of cast iron, bedded in clay, and upon fire-bricks. —Recollected the boiler of the Caledonia, London and Margate steam-packet, bursting at sea, by the forcing out of three of the rivets over the furnace-mouth, which extinguished the fire; but it was not productive of any injurious consequences to any of the persons on board; and the Cork and Cove packet-boat in Ireland, with two hundred and fifty officers and soldiers on board, burst her boiler when lying alongside of the transport that was receiving the troops. The bursting made a fissure, or opening, of nine inches by eighteen inches; but the steam which escaped did no injury either to the persons on board or to the vessel; nor does it appear, under any circumstances of the bursting of a wrought-iron boiler at the low pressure, the steam not being more than ten or fifteen pounds to the inch, that the steam which might be suddenly let loose or disengaged would have power sufficient to raise the deck of the vessel, or to injure the parties on board.—The Richmond steam-yacht cost, in the first instance, including the engine, 1800*l.* The engine itself cost about 1000*l.* The Majestic cost about 2000*l.* and the engine about 2000*l.* more. The Thames cost 2500*l.* including the engine at about 1200*l.* The new vessel built to go to Richmond, the hull and joiners' work cost 750*l.*; and an engine of fourteen-horse power and apparatus cost 1170*l.* The new Gravesend steam-yatch, the hull only has cost 750*l.* and the engine 1370*l.*; but there were various other expenses before these vessels could be finished.—Had just got a new boiler from Messrs. Jessop's, of Butterley, for the Thames steam-yacht, which was charged 215*l.*—A safety-valve would cost about 4*l.* and a mercurial tube for the same purpose 2*l.*—Had declined purchasing the Norwich steam-packet because it had a high-pressure engine.—Went with a party of German gentlemen from Bremen, who were anxious to make an immediate purchase of a steam-vessel; and they also declined to purchase that, or any of the boats upon the river Yare, solely because they had high-pressure steam engines on board.

Mr. Richard Wright, of Blackfriars Road, Engineer, called in and examined.

The boiler of the Norwich steam-vessel was eight feet long, with a cylindrical boiler four feet two inches diameter; it was first made with an internal angle iron at one end, and an external angle iron at the other end. In consequence of the internal angle iron having given way, a cast iron end was substituted, which certainly was not accurately performed. It was originally intended to sustain a pressure of forty pounds to the inch. —Should think that both wrought and cast-iron boilers might be used with equal safety; but that, in proving them, they ought to be kept under the pressure a considerable time, say a quarter of an hour, or half an hour. Sudden pressure may cause flaws in a boiler, which may give rise to accident afterwards; but, if under pressure a considerable time, the action of it might be seen.

Mr. JOHN RICHTER, of Cornwall Place, Sugar Refiner, called in and examined.

Was acquainted with the circumstances attending the explosion of the engine at the sugar house in Wellclose Square; and had attended from time to time, during the whole period of the construction of that boiler, for the purpose of boiling sugar by means of high-pressure steam; it was necessary they should have a pressure of from six and thirty to five and forty pounds to an inch.—Saw the boiler when the bottom only was put up, and was at that time informed that they had cast the dome part of it, and that it was not sufficient, and that they were casting another. Some months afterwards found that other placed there. Saw them at work; and was informed by Mr. Haigh, who was the engineer, that they were boiling at eighteen pounds an inch, but found the index of the gauge standing at five or six and thirty.—It was a mercurial gauge, intended as an index, and measuring inches. In consequence of complaints from Constant, the Frenchman, in whose house it was, that it would not do its work, and his fears in pressing it on to do its work, the maker of it became anxious to shew that it would, and a day was appointed for this to be done. Constant, at three o'clock in the morning, began his work, and continued boiling till about eight, but boiling with a great deal of difficulty, because he was afraid of putting the engine to the pressure he required. He gave it up; he said he would boil no more; and the men in attendance, who belonged to the engineer, went to fetch the engineer. He and his men came down, and persuaded Constant to have the fire lit again. He consented, after a great deal of difficulty, and went to another pan in an adjoining building, and there he was at work when the accident happened. They were urging the steam, and actually had put an immense weight upon the lever of the valve, so as to render it totally useless. This was ascertained by a Frenchman, who saw it, and who stated to the man that he was doing mischief, and doing wrong. He was told to hold his tongue, and mind his own business; that he knew his business, and they knew theirs: the consequence was, that immediately afterwards it blew up. After this accident went every day to the ruins, for the purpose of ascertaining what had been the cause of the bursting; and saw the excavation, until the parts of the boiler, which was of cast iron, were found, and then finding parts of this boiler in different places, the seat of the boiler being where it had been placed, but the rest scattered about in different directions. The bottom of it was two inches and a half thick, the upright sides of the bottom one inch and a half thick; the lower part of the dome was seven-sixteenths thick, and one of the parts at which it must have burst, and where the boiler was completely defective in the casting, was less than the eighth of an inch thick; it was not thicker than a crown-piece: the wonder is that it stood at all.— It was not intended to be worked above forty-five, and was ordered to be made to sustain the pressure of a hundred pounds to an inch. The whole house was blown to pieces, which arose from the fragments of the boiler striking the story posts, by which the support being taken away, the walls fell inwards.

Mr. John Steel, of Dartford, Engineer, called in and examined.

If it was required to make the strongest boiler imaginable, should consider cast iron preferable, because it could be got to an unlimited strength of resistance, while wrought iron could only be had of a certain thickness.—Was of opinion that the proof arising from the pressure of cold water was sufficient to ascertain the safety of a boiler, which should afterwards be exposed to the operation of fire, or of highly heated steam; cast or wrought iron being at its greatest strength at 300 degrees of heat, which had never been arrived at yet by steam.—Considered the mercurial gauge, and two safety-valves as essential in the construction of boilers; and was of opinion, that, by the adoption of those precautions, high-pressure steam might be used with safety, either with wrought-iron or cast-iron boilers.

Mr. William Brunton, of Birmingham, Civil Engineer, called in and examined.

Had been concerned in making boilers for high-pressure engines, which might be so constructed as to become useless before they were dangerous, upon the principle of having the exterior part of the boiler independent of the flue, so much so, that, while the flue is injured by the current action of the fire, the exterior part of the boiler remains, as to strength, unimpaired.—Conceived that a boiler thus formed, when the flue has been worn very thin, and then exposed to a greater pressure than it could sustain, the thin parts of the flue would act as so many safety-valves.—Believed it possible to construct boilers which would bear an expansive force of six hundred pounds to an inch.—Usually employed two safety-valves; one in an iron box under lock and key, and that only at the control of the proprietor, and the other open to the engine-man; and a mercurial gauge as an inverted syphon, which, in the event of the steam being stronger than the mercury can sustain, the mercury will be driven out, and the boiler thereby relieve itself.—In the high-pressure boiler, the injury which would arise from its bursting, would be done principally by the fragments projected; in the low-pressure boiler, the mischief might arise chiefly from the hot water and steam. Could mention two instances in illustration of this; the first of a low-pressure boiler having given way in the bottom, when a stream of hot water was projected against the engine-man, causing his death; the second instance was of a high-pressure boiler, in which a hole was suddenly opened, the water projected itself, and completely wetted a boy, standing within a yard of the orifice, who was not at all injured thereby. Should say the fragments from the cast-iron boiler would be equally destructive either with a high or with a low pressure. Considered that the fragments from a wrought-iron boiler would be projected with equal force with one of cast iron under equal circumstances.—Knew a wrought-iron boiler which burst with high-pressure steam; and a fragment, the largest piece, was carried to the distance of one hundred and fifty yards.—Was induced to prefer wrought to cast-iron boilers from the examination of several cast-iron boilers, which were cracked or broken in the lower part of them, which appeared to arise from the unequal temperature and expansion in the exterior part of the boiler, which was caused by a quantity of water at all times under the flue, and consequently of lower temperature than the water above the flue; thereby causing the upper part of the boiler to expand in a greater ratio than the under part of the boiler.—For steam navigation, would recommend a wrought-iron boiler, if properly constructed, and, at least, two safety-valves; the one to be placed under the lock and key of the proprietor of the vessel, so secured as not to be accessible to the engine-man; and one over which the engine-man had the usual control.—Would recommend the valves to be nearly flat, or quite so, as they would be less liable to be fastened by the difference of temperature to which the valve and the seat might occasionally be subjected.

Mr. George Dodd again called in and examined.

Had been on board and was well acquainted with twenty steam-boats; knows that there are more than forty in Great Britain; many of which had cost 5,000*l.* others 6,000*l.* and one on the Thames above 10,000*l.*; considered a fair average to be 3,500*l.* each, making the vested capital 140,000*l.* Most of them were fitted up with peculiar elegance and accommodation, the furniture and decorations alone forming an expensive item; they were also very expensive to maintain, especially on the Thames, by reason of the great cost of coal. They were most numerous on the Clyde, where they had been productive of essential benefit to the general commerce and traffic of Glasgow, Port Glasgow, Greenock, and the neighbouring country.—All of them had low-pressure condensing engines, and wrought sheet-iron rivetted boilers, except the remaining steam-boats between Yarmouth and Norwich, and one in Holland, built at Yarmouth; and they were high-pressure engines.

Mr. Josias Jessop, of the Adelphi, Civil Engineer, called in and examined.

Had no doubt but what the low-pressure boiler was more secure than the high-pressure, yet, from the natural wear and tear, both were liable to accidents. If an accident happened to one of a high-pressure, its consequences certainly would be more dangerous than that of a low-pressure engine.—Thought that to ensure safety, the boiler should be able to withstand the proof of two or three times the pressure to which it was afterwards likely to be put, or rather the pressure to which it should be limited; if, for instance, it was meant to work it at fifty pounds' pressure, and it stood the proof of one hundred and fifty pounds, the presumption would be that it was secure; but, in the course of two or three years, any boiler would wear out.—Would recommend an additional safety-valve, to which the person working the engine should not have access.—Preferred malleable iron or copper for boilers, because it would not burst by an explosion as brittle metal would; it would probably rend at the joints.—Was of opinion that the boiler should be adapted to the shape of the boat; and that being taken for granted, the safety would depend upon the strength of the metal, and not upon the form. It should be made of such strength, that any indenture would not affect it. Although the form approaching to cylindrical was of course stronger than any other form, that which approached nearest to a sphere was the strongest, but a cylinder with hemispherical ends was best.

Mr. ALEXANDER NIMMO, of Dublin, Civil Engineer, called in and examined.

Was of opinion that the best form for the safety-valve was that of an hemispherical cup, with its convex surface downwards, resting upon a collar, and to the bottom of the cup a weight was to be hung, which had previously been adjusted; by this means, the valve was always steam-tight in every position, yet without danger of adhering, and must be lifted by the steam when it exceeded a given pressure; but the valve might also be lifted by a chain attached to its upper side, which was inclosed within the iron case, and might be drawn up by the engine-man, or any person on board, and which did not allow him to keep it down, or to confine it. Had also found it necessary to prevent the accumulation of water upon the top of this valve, arising from the condensed steam, when escaping; this was done by a small waste-pipe descending from the bottom of the pipe which conveyed away the waste steam. Had thought it advisable to make the steam-valves large, that the weight which was laid on, being of itself large, might easily admit of addition. Employed two boilers communicating, and two safety-valves; and a mercurial gauge, provided with receivers, so as to prevent the loss of the mercury in case of any sudden collapsation or disengagement of steam, also a tube of glass attached to the boiler, which exhibited the level of the water in the boiler, and precluded any idea of danger in the minds of the passengers.—Was of opinion that the construction of the cast-iron boiler admitted of its being made of wrought iron with equal strength; then the explosion of the cast iron one would be more dangerous, as it would fly in pieces, whereas the other would probably tear. It was scarcely possible to form cast iron everywhere equally strong, and if a part be weaker than the rest, either on purpose or by accident, that would not have the safety that would be obtained by a wrought-iron boiler; for instance, in cast-iron boilers, it was common to have holes, and if these were filled with some metal of different melting temperature from cast iron, more fusible for instance than that, the juncture would part first, and it might be made to tear as a wrought-iron boiler would do; and again, the wrought iron was so much more liable to oxydation than cast iron, that although found very efficient at first, its strength and tenacity might be very speedily altered; for these reasons, cast-iron boilers had been preferred where high-pressure engines were used; and, in small tubes, the tenacity of cast iron could be made greatly to exceed that which could be given to wrought iron in the same form.

Mr. Arthur Woolf, of Pool, in Cornwall, Civil Engineer, called in and examined.

Approved of the cast-iron boilers in preference to any mixture of metals, particularly those composed of a number of tubes; it being always necessary in boilers to have a certain quantity of surface exposed to the action of the fire, to contain heat and steam; and if that were done in one vessel, of course it must be of considerable size greater in diameter than if composed of a number of tubes; and the risk of explosion is in proportion to its quantity of surface.—Considered his patent boilers calculated for every purpose; they were generally adapted to high-pressure steam; his patent was taken out for a safe boiler for a high-pressure engine; indeed, in his own engines, he did not work the steam to that height as was done in what were called the high-pressure engines, as the novelty of his engine was that it worked the steam twice over.—Made his boilers to stand from fourteen to twenty times the pressure he ever made use of, and employed two safety-valves.—Did not think that the wrought-iron boiler would separate into so many pieces as the cast-iron boiler, but had no hesitation in saying, that cast-iron boilers were safer than wrought-iron boilers.—Could make a cast-iron boiler stronger and more to be depended on for great pressure than wrought-iron; but where great pressure was not wanted, wrought-iron could be made sufficiently strong to depend on; and was of opinion, that as great a number of accidents had happened from the bursting of wrought-iron boilers as from cast ones.

Mr. ANDREW VIVIAN, Miner and Engineer, of Cambourne, in Cornwall, called in and examined.

Considered that the danger attendant on working steam engines arose from making the steam-vessel of insufficient strength for the steam; every engineer ought to be well acquainted with the power of the steam, and make the steam-vessels in proportion to the strength of the steam required.—Recommended the use of not less than two safety-valves on every boiler where a high pressure of steam was required, and that the boilers be made of sufficient strength, and proved before used.—To prove the boiler, it was first necessary to fill it with water, loading the safety-valves with ten or twelve times the weight required for the engine, and then by injecting water into them, so as to lift those valves with ten times the weight required.—Conceived that a boiler so proved, and furnished with safety-valves, properly adjusted to its contents, perfectly safe in working with steam, whether high or low pressure.—Was accustomed to load the engines in the mines under his direction, to about forty pounds an inch; and the valves were then loaded to about forty-five pounds.—Thinks it very possible to lock up one of the valves, which may be so constructed as not to be liable to accidents from explosion.—Did not see any reason why, in any situation whatever, the use of an engine should be limited to the low pressure, or that which is usually called the condensing engine.—Conceived that cast iron could be made much stronger than wrought iron, with less difficulty; some of the cast-iron boilers being made two inches thick; and to make a wrought-iron boiler equally strong as that, would be very difficult to be accomplished by workmen.—Had known of no accident with high-pressure steam and cast-iron boilers; but had known an accident happen working with Boulton and Watt's low-pressure engine, which was on the 28th of November, 1811, in Wheal Abraham mine; a wrought-iron boiler, working with low-pressure steam, exploded there, and scalded six men, three of whom died of the burns they received in the course of a week afterwards. —Did not recollect any instance in which a wrought-iron boiler exploded, so as that any persons were killed by the fragments.—Did not conceive that water could issue to any great distance from a high-pressure boiler, as it must soon be steam.—Had never known any persons scalded by the steam or the water issuing from a high-pressure boiler; but remembered many instances of persons being scalded from the same cause by a low-pressure engine, only one of which came directly under his own eye.—Was quite of opinion, that boilers made of wrought iron for high-pressure engines would soon become leaky, and that too without exploding. Knew an instance of a boiler of that description made, which became leaky and unfit for use in a very short time; the consequence of which was, the working of the mine was stopped, and a great number of people thrown out of employ.—Supposing the only object to be safety to the lives or limbs of the persons who should be surrounding the engine, would, in that case, prefer having the boiler of a high-pressure engine of cast iron, because it could certainly be made stronger than wrought iron for the same expense; while he considered the risk was so small as that it scarcely need be taken into the question, because all explosions might be easily prevented by proving the boiler every time it was cleansed, which he thought should be at least every month.—Had found the use of a high-pressure engine of great advantage to the Cornish mines, which could be proved by the monthly reports.—Conceived that every engine ought to have two safety-valves, and one should be locked up to prevent careless engine-men doing mischief, which low-pressure engines are as liable to as high.—Was of opinion that a high-pressure engine did greater duty with the same coals than a low, which could also be proved by the monthly reports. — Being desired to attend the Hon. Committee on the part of the proprietors of three of the largest mines in Cornwall, the united mines of Crowan, Dolcoath, and Weal Unity, they wished to state their hope, that the Legislature would not interfere to prevent the use of high-pressure engines, either on board boats, or in any other way.

Mr. THOMAS LEAN, Inspector of Steam engines, of Crowan, in Cornwall, called in and examined.

Was employed by nearly the whole of the miners in Cornwall to inspect their engines, and make monthly reports of the work they performed. —Conceived there was no danger whatever in the use of high-pressure steam engines; and for this reason, that, in general, for an engine intended to be worked with high steam, the materials were made stronger in proportion than the materials used for steam of low-pressure.—Considered it of importance that every boiler should have two safety-valves, one of which should be confined from the engine-man.—In a boiler in which great strength was required, would certainly recommend cast iron, and had no doubt but it could be made much stronger than wrought iron, the explosions that had happened in Cornwall having all been in wrought-iron boilers, and from low-pressure steam.—In every boiler that was built, there was one part of it weaker than another, and it was hardly possible for a boiler to be thrown about in fragments to do mischief. Should not feel any hesitation to sit on the cast-iron boilers in Cornwall when an explosion took place, being convinced the explosion would take place at the under part.—Was in the habit of working the high-pressure boiler at forty pounds to an inch, while they were proved to three hundred, and that too without injuring the boiler.—Apprehended, that with a boiler so constructed, so proved, and guarded by two safety-valves, there would be no danger whatever in any situation; and was also of opinion, that the high-pressure engines in Cornwall had saved at least two-fifths of the whole consumption of coals in the county; in some instances it had saved three-fifths.

Mr. George Dodd again called in and examined.

Witness wished to offer to the Committee a second safety-valve, which admitted of being locked up so as to be inaccessible to the engineer. This was furnished with a flat bottom, resting upon a flat circular ring; the steam escaping from the sides of the box through apertures, so constructed as that nothing could be introduced to impede its action.

William Lester, Esq. of Lambeth, called in and examined.

Witness attended for the purpose of delivering in the drawing of a valve so constructed, as to prevent the possibility of any person having access to it to prevent its action; it was self-acting entirely from the gravity of a column of water acting upon the valve, which prevented its being locked by any mode, and it could not adhere because it was not a cone acting in another cone, but a flat surface pressing upon the top of a cylinder; and being enclosed in a box, and the steam getting out at the bottom, no matter could get upon the valve to cause its adhesion.

REPORT.

The Select Committee appointed to consider of the means of preventing the mischief of Explosion from happening on board Steam-Boats, to the danger or destruction of his Majesty's Subjects on board such Boats; and who were empowered to report their observations and opinion thereupon to the House; together with the Minutes of the Evidence taken before them; have, pursuant to the Order of the House, considered the matters to them referred, and agreed to the following Report:—

Your Committee entered on the task assigned them, with a strong feeling of the inexpediency of legislative interference with the management of private concerns or property, farther than the public safety should demand, and more especially with the exertions of that mechanical skill and ingenuity, in which the artists of this country are so pre-eminent, by which the labour of man has been greatly abridged, the manufactures of the country carried to an unrivalled perfection, and its commerce extended over the whole world.

Among these, it is impossible for a moment to overlook the introduction of steam as a most powerful agent, of almost universal application, and of such utility, that but for its assistance, a very large portion of the workmen employed in an extensive mineral district of this kingdom, would be deprived of their subsistence.

A reference to the evidence taken before your Committee, will also shew with what advantage this power has lately been applied, in Great Britain, to propel vessels both of burden and passage; how much more extensively it has been used in America, and of what farther application it is certainly capable, if it may not be said to be even now anticipated in prospect.

Such considerations have rendered your Committee still more averse than when they entered on the inquiry, to propose to the House the adoption of any legislative measure, by which the science and ingenuity of our artists might even appear to be fettered or discouraged.

But they apprehend that a consideration of what is due to public safety, has on several occasions established the principle, that where that safety may be endangered by ignorance, avarice, or inattention, against which individuals are unable, either from the want of knowledge, or of the power to protect themselves, it becomes the duty of Parliament to interpose.

In illustration of this principle, many instances might be given; the enactments, respecting party-walls in building, the qualification of physicians, pilots, &c. the regulations respecting stage-coaches, &c. seem all to be grounded upon it. And your Committee are of opinion, that its operation may, with at least equal propriety, be extended to the present case, on account of the disastrous consequences likely to ensue from the explosion of the boiler of a steam engine in a passage-vessel, and that the causes by which such accidents have generally been produced, have neither been discoverable by the skill, nor controllable by the power of the passengers, even where they have been open to observation.

Your Committee find it to be the universal opinion of all persons conversant in such subjects, that steam-engines of some construction may be applied with perfect security, even to passage-vessels; and they generally agree, though with some exceptions, that those called high-pressure engines, may be safely used with the precaution of well-constructed boilers, and properly adapted safety-valves; and further, a great majority of opinions lean to boilers of wrought iron or metal, in preference to cast iron.

Your Committee, therefore, in consequence, have come to the following resolutions, which they propose to the consideration of the House:

1. Resolved, That it appears to this Committee, from the evidence of several experienced engineers, examined before them, that the explosion in the steam-packet at Norwich, was caused not

only by the improper construction and materials of the boiler, but the safety-valve connected with it having been overloaded; by which the expansive force of the steam was raised to a degree of pressure, beyond that which the boiler was calculated to sustain.

2. Resolved, That it appears to this Committee, that in the instances of similar explosions, in steam-packets, manufactories, and other works where steam-engines were employed, these accidents were attributable to one or other of the causes above alluded to.

3. Resolved, That it is the opinion of this Committee, that, for the prevention of such accidents in future, the means are simple and easy, and not likely to be attended with any inconveniences to the proprietors of steam-packets, nor with any such additional expense as can either be injurious to the owners, or tend to prevent the increase of such establishments. The means which your Committee would recommend, are comprised in the following regulations:

That all steam-packets carrying passengers for hire, should be registered at the port nearest to the place from or to which they proceed:—That all boilers belonging to the engines by which such vessels shall be worked, should be composed of wrought iron or copper:—That every boiler on board such steam-packet should, previous to the packet being used for the conveyance of passengers, be submitted to the inspection of a skilful engineer, or other person conversant with the subject, who should ascertain, by trial, the strength of such boiler, and should certify his opinion of its sufficient strength, and of the security with which it might be employed to the extent proposed :—That every such boiler should be provided with two sufficient safety-valves, one of which should be inaccessible to the engineman, and the other accessible both to him and to the persons on board the packet :—That the inspector shall examine such safety-valves, and shall certify what is the pressure at which such safety-valves shall open, which pressure shall not exceed one-third of that by which the boiler has been proved, nor one-sixth of that which by calculation it shall be reckoned able to sustain :—That a penalty shall be inflicted on any person placing additional weight on either of the safety-valves.

4. Resolved, That the Chairman be directed to move the House, that leave be given to bring in a Bill for enforcing such regulations as may be necessary for the better management of steam-packets, and for the security of his Majesty's subjects who may be passengers therein.

CHAPTER V.

Cylinder and Piston — Condenser — Air-Pump — Barometer and Steam-Gauge—Working Beam— Parallel Motion — Counter — Sun and Planet Wheel—Fly and other Modes of regulating Velocity—Boiler—Safety-Valves—Furnace.

HAVING taken a brief review of the early history and general principle of this stupendous machine, it may be advisable before we proceed to a description of the principal engines now employed, to examine more minutely the separate parts and the progressive improvements effected in each.

The cylinder and piston being those parts of the engine in which the effective force is more immediately produced, may first claim attention.

The piston of the atmospheric engine is generally made of cast iron nearly fitting the inside of the cylinder, a circular ledge or rim being formed round it to receive the packing, without which the steam would find a passage through the interstices in the cylinder. Mr. Smeaton, who greatly improved the atmospheric engine, coated the under side of the piston with elm or beech planks about two inches and a quarter thick; the wooden bottom being screwed to the iron with a double thickness of flannel and tar, to exclude the air between the iron and the wood. By the adoption of this improvement its property of conducting heat was reduced, and the wood having been previously jointed with the grain radiating in all directions from the centre, was not liable to expand by the heated steam. This piston was kept air-tight by a small stream of water continually falling on its upper surface; but in Mr. Watt's engine he was compelled to effect this by improving the fitting of the piston, the old mode being inadmissible. It is now cast with a projecting rim at bottom, which is fitted as accurately to the cylinder as it can be, to leave it at full liberty to rise and fall through the whole length. The part of the piston above the rim is about two inches less all round than the cylinder, to leave a circular groove for the hemp which forms the packing. To keep this in its place, a lid or cover is put over the top of the piston, with a ring or projecting part, which enters into the circular groove for the packing, and pressing upon it the plate is forced down by screws, which work into the body of the piston. By this means the packing is made to fill the diameter of the cylinder with tolerable accuracy, and to prevent for a time any steam passing between the piston and the cylinder. When, however, by continued working the piston became too easy, and so occasioned a waste of steam, it was found necessary to take off the top of the cylinder to get at the screws, even when fresh hemp or packing was not wanted, and this operation being attended with considerable labour, was seldom resorted to by the engine-man till a great waste of steam had taken place. By an improvement on this piston introduced by Mr. Woolf, this is now effected without taking off the cylinder cover, except, indeed, when new packing is required.

To accomplish this, Mr. Woolf fastens on the head of each of the screws a small cog-wheel or nut, and these are all connected together by means of a central wheel working loose upon the piston-rod in such a manner, that if any one of the screws be turned a similar motion is given to the remainder, a cap being provided in the upper end of the cylinder screwed down by bolts to make it steam tight. In a piston thus constructed, there is little difficulty in drawing down the packing, by applying a key to the square head of the projecting screw employed to communicate with the rest. Another method contrived by Mr. Woolf for the smaller pistons differs but little from the preceding in construction. Instead of having several screws all worked down by one motion, there is in this but one screw, and that one cut upon the piston-rod itself; on this is placed a wheel, the centre of which is furnished with a female screw, which is forced down by means of a pinion furnished with a square projecting head turned in a similar manner to the preceding.

For high-pressure engines, however, the metallic

piston invented by Mr. Cartwright has the most decided preference. This not only saves the trouble and expense of packing, which must be frequently renewed in all other engines, but also a great deal of steam, on account of the more accurate manner in which it is made to fit the cylinder; this is effected in the following manner: Two metal rings are accurately ground into the cylinder, so that no steam can pass between their exterior surface and the inside of the cylinder, their upper and under sides are also ground perfectly flat, and applied one upon the other. On the upper ring is placed a plate of metal, rather smaller in diameter than the cylinder, while a similar flat plate is placed below the under ring, both of which, with the rings between, are attached firmly to each other by means of the piston-rod passing through them.

A shell being thus formed, the rings are each of them cut into three pieces, and in cutting them, such a portion of the metal is taken away as to leave room to introduce between two of the pieces, a spring in form of the letter V, the open end of which is placed outwards, almost close to the circumference; by which means the two pieces against which the two sides of the spring act, are pressed in the direction of the circumference, against the ends of the third piece, so that the three pieces are thus kept so uniformly in contact with the cylinder, that the longer the machine is worked, the better the rings must fit. To prevent steam passing through the cuts in the lower rings, the solid parts in those upon the upper side, are made to fall upon the divisions and springs of the under ones, thus interrupting the communication that would otherwise remain open, and forming a perfect break-joint. The interior surface of the cylinder in which the piston works, requires to be bored with the greatest exactness, though this was but little attended to in the early atmospheric engines, some of them being composed of timber hooped together in the same manner as barrels are constructed. Mr. Watt, in his first attempts at improving the steam engine, employed this material in the construction of his cylinders, though he afterwards abandoned it for those of bored metal; the operation of boring being performed with the greatest precision, by an apparatus invented by Mr. Wilkinson*.

Mr. Murray has also effected considerable improvements in this part of the engine, and the boring machines employed in his manufactory are of considerable value. They are worked by a separate steam engine, which is never stopped during the operation, as in that case a shoulder or ring would be formed, running completely round the cylinder.

In small engines, it is common to place the cylinder within the boiler, in which case no artificial

---

* For a description of Mr. Wilkinson's patent cylinder apparatus, see Appendix, A.

mode of retaining the heat is required; but to this arrangement in those of larger dimensions there are several objections, not the least of which is the frequent repairs that are necessary in the boiler; and a similar effect has been produced by the use of a double cylinder. This was first adopted by Messrs. Boulton and Watt, the outer cylinder or steam-jacket keeping the inner cylinder at the temperature of boiling water, by the action of a partition of steam made to pass between the jacket and the working cylinder.

We have already stated, that Mr. Watt's great improvement consisted in condensing the steam in a separate vessel where a vacuum was formed by the continued application of cold water. A metal box constructed for this purpose, and furnished with a pump for drawing off the water and air, is called a condenser. It is necessary that the parts appropriated to this purpose should be kept as cold as possible; and upon this account the air-pump and condenser are placed in a cistern of cold water, which is kept full by the continued action of a pump, also worked by the engine, and called the cold water pump, a little being allowed to pass off continually to preserve the water at an equable temperature.

The air-pump and condenser are usually of the same size; if of one eighth, the capacity of the working cylinder, it will be found sufficient to keep the condenser empty in Mr. Watt's single engine. The best proportion for a double action engine is about two-thirds the diameter of the cylinder and half the length of stroke, the condenser, as in the single engine, being of similar capacity.

In Mr. Maudslay's portable engine the condenser is an hollow cylinder, and the air-pump is placed within it, so that there is no necessity for a pipe of communication from the air-pump to the condenser; and in this case a small cistern is fixed over the pump to contain the hot water, the discharge-valves being placed in the lid, which thus forms the bottom of the cistern or hot well.

In the early engines, on Messrs. Boulton and Watt's construction, the air-pump and condensing-cistern were placed at the outer end of the beam; in which case the pump-bucket being drawn up by the descent of the piston, the engine required less counter-weight than in the present form, in which the air-pump must be wholly worked by the counter-weight. It was necessary also, that the parts appropriated to the condensation of steam should be kept as cold as possible; on which account, the air-pump and condenser were placed in a cistern of cold water, which being continually on the overflow, carried off the excess of heat.

The mode of condensing by outward cold, was not however found sufficient; and Mr. Watt afterwards introduced a small jet of water, the dimensions of the air-pump being so far increased, as to extract the injection-water as well as the air.

To shew the degree of vacuum in the condenser, and consequently the amount of pressure on the piston, a barometer-gauge has been employed. This is justly considered as a most important instrument, though unfortunately for the profit of steam engine proprietors but little attended to. This gauge is in fact a common barometer tube, of thirty inches in length, with a graduated scale, and connected with the condenser by a small tube furnished with a stop-cock. When the air is expelled from the cylinder this must be closed, otherwise the steam entering the tube would blow the mercury from the cup. On the cock being turned, and the communication opened with the condenser, the exact degree of vacuum will be shewn by the height of the mercurial column, which, if the condensation be not complete, or air be admitted, will descend, and on the contrary, if perfect, it will ascend, as in the Torricellian tube.

The steam-gauge employed by Mr. Watt, consists of an inverted syphon or bent tube of glass or iron, one leg of which is jointed to the steam-pipe, while the other is open to the atmosphere. A quantity of mercury being poured into the tube, it will occupy the lower or bent part, and the surface of the fluid metal in one leg being exposed to the pressure of the steam, while the external air acts upon the other, it is evident that the difference of level of the two surfaces will express the pressure of the steam in the height of the mercurial column passing up the graduated tube.

This gauge is just the reverse of the preceding; the barometer shewing the pressure of the atmosphere on a given space of the piston, while the steam-guage indicates the force of elastic vapour entering from the boiler. It is the duty of the fire-man frequently to look at this gauge, that he may know when to increase the fire in the furnace, and thus a sufficient supply of steam will always be secured to the engine.

In the early atmospheric engines, the working-beam was composed of a large and almost unhewn tree; but Mr. Smeaton employed a framing of wood for this purpose, which was afterwards much simplified and improved by Mr. Hornblower.

In double-acting engines it is usual to have the beam cast in one piece, the extremities being turned in a lathe to form cylindrical pins, upon which are fitted sockets or pieces, having other pins projecting from them to form the points of the parallel-motion and connecting-rod. Thus, there is one pin on each side of the socket, the two links of the parallel-motion being fitted to the two projecting pins at one end, while the double joint of the connecting-rod is fitted on the two pins at the other end of the beam. The advantage of this construction is, that the joints at the ends of the beam become universal joints, having liberty of motion in all directions; and in some of Mr. Murray's best engines, the same contrivance is applied to the crank-pin and connecting-rod.

The two great links of the parallel-motion, are each composed of a strap or loop of iron, bent so as to form a double link, in the upper bend of which are two sockets for the pivots at the end of the beam, and at the lower end are two others, for the pivots which project on each side of the piston-rod socket. The brasses of this joint are held in by wedges, put through the two links at the lower end, which, on being driven inwards, tighten the fittings at pleasure.

To ascertain the number of strokes made by the engine in a given time, a simple apparatus was contrived by Mr. Watt, called a counter. This is in some cases attached to the beam, each stroke moving one tooth, and the index hand shews how many strokes have been made in a given time; and by comparing this register with the diameter of the piston, and the barometer-gauge, the exact power of the engine is accurately shewn.

The fly-wheel has justly been considered one of the most important and valuable parts of the steam-engine: when combined with the crank, it is employed to convert a reciprocating into a rotatory motion. If of moderate size, it should be cast in one piece of metal; this, however, cannot often be accomplished from its great weight, the fly-wheel of a large engine frequently exceeding ten tons. When of this size, the ring is usually cast in six pieces, of about a ton each, and connected by wrought-iron bolts; but a method has lately been introduced in large engines, of substituting the dove-tail for that mode of connecting the parts. In this case the arms are fastened into the ring, and the segments of the ring fastened together by a system of dove-tails, which admit of being put together only in one direction, which is contrary to that in which the centrifugal force acts. It is a great object in constructing fly-wheels, to choose that form which offers the least possible resistance to the medium through which it revolves, and on this account the ring should be smooth and truly circular; the radii being made with a thin edge to the air. It is also necessary that the various pieces connected with the fly, should be cast in the most solid manner, as the centrifugal force of so large a mass frequently moving at the rate of more than three hundred feet per second, would, in the event of any part flying off, be productive of the most fatal consequences.

Messrs. Murray and Wood form the radiating arms or cross bars of an elliptic figure, the narrowest edge meeting the air; and to these eminent engineers we are indebted for the following rule for proportioning the fly-wheel of the steam engine. Multiply the number of horse power of the engine by 2000, and divide it by the square of the intended velocity of the circumference of the fly-wheel in hundred-weights. Of this rule Mr. Buchanan furnishes an example: to find the weight of a fly-wheel proper for an engine of twenty horses' power, supposing the fly-wheel to be 18 feet in diameter, and to make 22 revolutions per second: wheel 18 feet diameter = 56 feet circumference; × 22 revolutions per minute = 1232 feet motion per minute ÷ 60 = $20\frac{1}{2}$ feet motion per second for the motion of the circumference of the fly-wheel. Then $20\frac{1}{2}$ feet per minute squared, = $420\frac{1}{4}$, and twenty horses' power × 2000 = 40000 ÷ $420\frac{1}{4}$ = 90·4 cwt. of the wheel required.

In addition however to this mode of regulating the velocity of the steam engine, a variety of plans have been suggested for equalizing the admission of steam; the most simple of which is by means of a handle connected with the throttle-valve. This is a thin circular vane placed in the steam pipe, turning on a pivot across its centre, which comes through the pipe, and has a small handle fixed on the end of it, by turning which, the passage is opened or shut. When the vane is set, so that its plane is perpendicular to the axis of the pipe, it nearly fills the circular passage, and allows very little steam, if any, to pass by it; but when the vane is turned edgeways, it presents a very small surface, and leaves the passage nearly open; so that by thus turning the handle, the attendant can at any time regulate the speed of the engine.

The governor or double pendulum, is also employed for this purpose. This consists of two balls, suspended by joints projecting from a vertical axis, which being caused to revolve by the machine to which it is connected, will increase the diameter of the path described by the balls with the increasing speed of the machine; or, in other words, their centrifugal force will cause them to fly off from the arbor in a degree proportionate to the velocity of the machine; and this motion is made to actuate the lever connected with the valve, which admits the steam from the boiler to the cylinder.

Another method, is to have the small pump worked by the engine, and raising up water into a cistern, from which it runs out again in a constant stream. By this means the water will accumulate, and rise in the cistern, if the engine work rapidly, so as to pump more water into the cistern than will flow out of it in the same time; and, on the contrary, the surface of the water will sink in the cistern, if the engine work slowly, and a float being in the cistern, and connected with a wire to the throttle-valve, a proportionate effect will be produced on the engine.*

---

* The patent regulator invented by Mr. Job Rider, is described in Appendix, A.

The only materials that have hitherto been employed in the construction of steam-engine boilers, are iron, copper, wood, and stone. The latter of these was introduced by Mr. Brindley, who, in 1756, erected a steam engine near Newcastle-under-Lyne with a boiler of this description. It was composed of brick and stone firmly cemented together, and the water was heated by iron flues passing in various directions. An admirable cement for this species of boiler may be formed of boiled linseed oil, litharge, and red and white lead mixed together to a proper consistence, and then applied as a species of mortar to the stones. If the joints be properly filled, a cistern thus constructed will never leak, nor want any very considerable repair.

Savery's boilers were of copper, and contained about five or six gallons of water; and the Marquis of Worcester states, that he employed "a piece of a whole cannon" for that purpose.

The atmospheric-engine, constructed by Newcoman, was provided with a boiler of considerable dimensions, composed of wrought iron plates, the upper part being of an hemispherical form to resist the elasticity of the steam; and it is of considerable importance that this part of the boiler be accurately proportioned to the power of the engine. If the boiler-top be too small, it requires the steam to be heated to a greater degree to increase its elastic force sufficiently to work the engine, and then the condensation on entering the cylinder will be greater. If the top contain eight or ten times the quantity of steam used at each stroke, it will require no more fire to preserve its elasticity than is sufficient to keep the water in a proper state of boiling; this, therefore, may be considered as the most eligible size.

Wooden boilers have as yet we believe been exclusively confined to America. They were introduced by Mr. Anderson and Chancellor Livingston. The merits of this boiler are, economy in construction, and a very material saving in fuel; the latter of which advantages will be readily seen from the circumstance, that wood is a bad conductor of heat, while metal is one of the best. That there is a great saving in the employment of this species of boiler where wood is cheap is sufficiently evident; that part, however, which is above the water, and consequently exposed to the action of the steam, speedily decays, and the elastic vapour passes through the joints. This, however, might be remedied, by coating the internal surface with thin metal, which might readily be connected with the furnace and flue, so as to make the whole boiler steam-tight.

The boiler of Messrs. Boulton and Watt's engine is so placed as to receive the greatest possible degree of heat, the flame passing through a long flue which twice encircles the lower part. This is kept constantly supplied with water, to repair the waste of evaporation by means of a pump communicating with the hot-well; and as it is necessary that this should always be preserved at the same level, the feed-pipe is closed by a valve in the bottom of the cistern, which prevents the water running down into the boiler until its level subside, and shews that it requires replenishing. To know the exact height of the water in the boiler, two cocks are mostly employed; one of which is carried below the requisite water-mark, and the other stands a little above the desired point. If water should issue from both cocks, the supply has not been sufficient, and more must be admitted; but if, on the contrary, water proceed from the one cock and steam from the other, it may then be considered about the proper level.

The patent boiler employed by Mr. Woolf, is different from that commonly used in engines which work with steam of a low pressure, the water being contained in several cylindrical tubes of cast iron which are exposed to the heat of the furnace nearly in an horizontal position. In the employment of this kind of boiler care should be taken that the flame and heated air be made to come completely in contact with the iron tubes of which it is composed, and so as to give out the least possible portion of heat previous to reaching the chimney.

This mode of raising steam of great elasticity, by exposing a large surface in a number of heated tubes, does not appear to have originated with Mr. Woolf, it having been proposed by Mr.

Blakey in a small tract which he published in Holland as far back as 1776. It appears, however, that Blakey's tubes were to be placed over each other upon the same principle as the olefient gas-retorts, and the water passing down the heated pipes, was thus readily converted into steam.

The high-pressure boiler, employed Mr. Trevithick, is supplied with water previously heated in a separate vessel, by a small force-pump worked by the engine. In some of the improved engines, however, another and more ingenious mode had been adopted, the water being driven in by the action of a volume of highly expansive steam.

Among the provisions made for the security of the high-pressure boiler, we may enumerate the soft metal plug and double safety-valve. The former of these contrivances is calculated to prevent the boiler being burst by the sudden introduction of water, when it has been allowed by carelessness to boil dry, and become red hot; and by the employment of two safety-valves placed in different parts of the boiler, the chance of accident is diminished at least one half, while the effect of the engine is in no shape impaired.

The greater part of the boilers employed in American steam-navigation are of wrought iron, and are usually more than a quarter of an inch in thickness, of a cylindrical form, and about thirty inches in diameter, with a cast-iron end about two inches thick; and the testimony of experienced engineers both in this country and America, has invariably shewn, that such a species of boiler, when old and thin by long wear, has generally given way by a small rent or fissure through which the steam escapes, gradually taking off the internal pressure, and thus securing the passengers adjacent from the dreadful consequences which have frequently resulted from the explosion of cast-iron boilers similarly constructed.*

The safety-valve being an object of considerable importance, both as regards the utility of the engine, and the preservation of those connected with its management, much attention has been given to its construction, and to this highly useful appendage we would particularly call the reader's attention.

The first engine that was made by Captain Savery had a steel-yard safety-valve, to let the steam fly off when it arrived at a dangerous degree of elasticity. The following figure will furnish a sufficiently accurate idea of this simple apparatus. A, the top of the boiler; B, the safety-valve or plug made to fit air tight in the tube or valve-seat beneath; C, the lever working on an axis at D, and furnished with a moveable weight, E, adjusted to balance the pressure of the steam.

---

* In those boilers that are constantly employed with sea-water, a great accumulation of salt takes place; it is therefore necessary for ships which perform long voyages to be provided with two boilers, each of which should be capable of supplying the engine with the necessary quantity of steam.

When steam of considerable elasticity is required, the weight is placed at the extremity of the lever, and as such acts with greater force on the safety-valve than when removed to a point nearer to the axis on which it revolves. So that should low-pressure steam be required, it will only be necessary to remove it nearer the axis or centre, and *vice versa.*

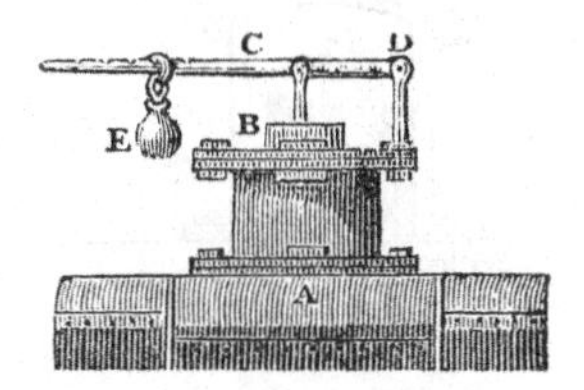

The lever and balance-ball which form this apparatus, would at all times be effectual were they not liable to be fastened by the corrosive nature of the materials of which the valve is composed, and, what is worse, their pressure altered by the addition of more weight. This, however, as too frequent experience has shewn, is continually the case, the engineer having more regard for the full performance of his machine than for his own safety or life; and to the overloading of this valve, these accidents may be principally attributed.

To prevent a recurrence of those accidents which first drew the attention of the legislature to this important part of the engine, and to which we have already referred, under the head of steam navigation, it appears advisable to inclose the safety-valve in an iron box, and so put it beyond the control of the engine-man.

The annexed figure represents an inaccessible safety valve, calculated to answer all the purposes for which it is intended, namely, the preservation of those employed in the neighbourhood of the boiler, and economy in the use of steam.

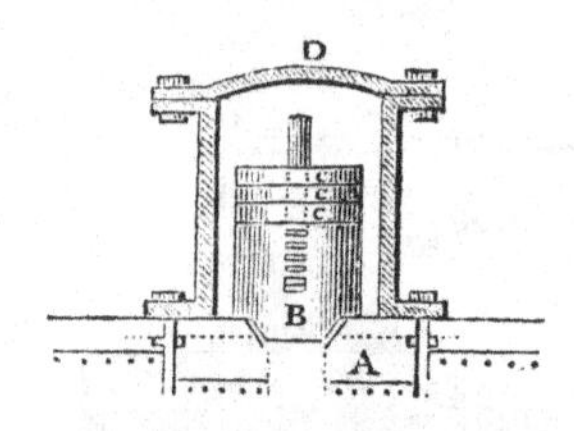

In this, as in the preceding diagram, A represents the boiler, and B the safety-valve, furnished with a small upright staff, on which slide the additional weights CCC. The whole is inclosed in a box D, pierced with holes to allow the steam to escape after it has raised the valve B.

Should high-pressure steam be wanted, it is necessary only to increase the number of weights, and the desired effect is produced; or if, on the contrary, steam of the usual atmospheric pressure be wanted, the whole of the weights are taken off.

The safety-valve invented by the Chevalier Edelcrantz, has nearly the same properties as that employed by Mr. Woolf. It consists of a small brass cylinder which is fixed on the boiler, and fitted with a piston made to descend with its own weight when raised by the pressure of the steam.

The lower part of the cylinder being made to communicate with the boiler; the upper part is closed by a small cover screwed on to it, and perforated with a hole, through which the piston-rod passes freely, which serves the double purpose of keeping the piston perpendicular, and preventing it being blown out. The sides of the cylinder are pierced with a number of small holes, placed in succession at a short distance above each other, so that the open space for the steam to escape, increases with the height of the valve, and is ultimately enlarged so as to prevent any danger of explosion. The piston-rod is also furnished with a number of weights, fitting loosely on a small shoulder, similar to those employed in the common hydrometer; and these may be removed or increased at pleasure.

Another advantage likely to result from the adoption of this safety-valve is, the facility with which it may be employed to regulate the fire of the steam-engine furnace to the intensity of the elastic vapour required. This may readily be effected by a register pressing on the top of the safety-piston, and connected with the apertures for the admission of air, which, by increasing or decreasing the supply of oxygen, will have a proportionate result on the steam generated in the boiler, and consequently effect a considerable saving in the expenditure of fuel.

Another safety-valve, opening internally, has, we believe, also been added by Messrs. Boulton and

Watt. This is of great utility, more particularly in large engines, as it prevents the sides of the boiler being crushed in by the sudden introduction of water, or any artificial condensation that may take place from reducing the heat of the boiler-head.

CHAPTER VI.

Savery's Engine improved by Pontifex.—Atmospheric Engine.—Single-acting Engine, by Boulton and Watt.—Murray and Wood's Engine.—High-pressure Engine.—Woolf's Double-cylinder Expansion Engine.—Maudslay's Portable Engine.—Masterman's Rotatory Engine.—Smoke-consuming Furnaces.

THE Engine invented by Savery, and improved by Pontifex, possesses considerable advantages over the Marquis of Worcester's apparatus, and it is probable that the extreme simplicity of this engine will, when better known, bring it into more general use. With this view we have selected it as the subject of our first plate, in preference to the original engine, the principle of which has been already very fully explained. The apparatus we are about to describe, has lately been erected at the City Gas Works.

Plate I. Fig. 1. and 2. represent front and side elevations of the cylinders, and connecting apparatus.

Fig. 3. Back view, with section through the cistern and buckets.

L

Fig. 4. and 5. Vertical and horizontal sections; the latter commencing at the dotted line *aa*, Fig. 1.

Fig. 6. and 7. Side and end view of the waggon boiler.

b.b. Fig. 1. Two steam cylinders connected by cross tubes at *cc*, in each of which a vacuum is alternately formed by the condensation of elastic vapour, conducted from the boiler by the bent-tube *d*, and admitted to the steam-cylinders by means of the sliding-valve *e*.

f.f. Fig. 4. Two tubes perforated with small holes for the admission of steam and injection water, the latter of which is distributed by falling on the strap *g*.

h. The suction-pipe proceeding to the bottom of the well, which in no case ought to exceed from twenty-eight to thirty feet in depth; so that a vacuum being formed in the copper vessels *b b*, the water will be raised by the pressure of the atmosphere, and passing up the tube *h* will take the place of the elastic vapour.

i. i. Two valves placed at the upper end of the suction-pipe *h*, which allow of the upper passage of the water from the well, but prevent its return.

j.j. Two similar valves opening into the air-vessel *k*, to which is attached the nozzle *l*, serving to convey the water from the copper vessels to any required point.

m. The injection tube, furnished with a valve

at *o*, and intended to convey water from the box *n*, to the taper tubes *f f*.

p. Stop-cock to regulate the supply of condensing water.

q. Tube passing from the bottom of the cistern *n*, to the injection-tube *m*, and furnished with a stop-cock at *s*:

The quantity of water in the cistern *r* is regulated by a floating valve *t*, which in Fig. 3. is represented immediately over the pipe *q*, so that the valve is opened whenever the water rises beyond the required depth.

u. u. Two tubes communicating with the back part of the chambers *n n*, and the inverted vessels *v v*, each tube being furnished with a valve at *w* to admit the water from the chamber to the buckets *x x*.

x. x. Two buckets suspended by rods, and a chain passing over the wheel 2, which is fixed on the end of the axis 3, and supported by a bracket 4. From the other end of the axis 3 projects an arm 6 provided with a stud T.

9. A horizontal axis turning in a stuffing-box at 10, on one end of which is fixed a pinion 11, which serves to give motion to the sliding-valve *e*.

To put this engine in action, the steam must be first raised to the boiling point, and the valve or cock opened, which admits it to pass from the boiler to the pipe *d*. One of the buckets must now be made to descend, which will open the sliding-valve *e*, and admit the steam into the cylinder *b*, 1 The atmospheric air, which will thus be expelled from the cylinder, is allowed to pass through the valve *j* and nozzle *l*. The other bucket must then be depressed, and by its action upon the sliding-valve it will open a communication for the injection water through the pipe *q q*, which passing down the perforated tube *f* will immediately condense the steam, and form a vacuum in the vessel. The whole pressure of the atmosphere being now removed from the suction pipe *h*, the water will rush up to restore the equilibrium, and the vessel *b* being filled will furnish a supply at the bent-tube *l*.

Having examined the action of one-half of the apparatus, we may suppose the same effect to be produced on the opposite side. The steam will, in the first instance, be admitted by the pipe *c*, and a communication afterwards opened by means of the sliding-valve with the condensing water, which by reducing the steam to its original bulk will form a vacuum, and the water will again ascend as in the first vessel.

The stop-cock *y* must now be opened, and the bucket *x* first described made to descend, which will remove the sliding-valve *e* to its original position, and admit the steam to the upper part of the first vessel, which will depress the water, and cause it to flow through the valve *j* and nozzle *l*, while at the same time the water will pass through the tube *u u*, in which the valve *w* is inserted beneath the inverted vessel *v*. The water will continue to enter the bucket *x* till its increasing weight causes it to preponderate, and turn the sliding-valve *e* in the opposite direction.

Should there not be a sufficient supply of water in the cistern *r r* for the purpose of condensing the steam in the large vessels, the stop-cock *p* must be opened, and an additional supply of water will then be furnished from the chambers *n n* by the tube *m*, and in the event of the bucket not being depressed at the instant that the water is expelled from the chamber *n* of the vessel *b*, the steam will pass through the tube *u u*, and act between the under side of the fixed inverted vessel *v*, and the surface of the water in the moveable bucket *x*, the descent of the bucket being accelerated by the repellant force of the steam, so that by the alternate action of the buckets *x x*, the motion of the engine is rendered continuous.

It appears that each steam-vessel in the engine employed at the City Gas Works, contains about thirty-six gallons of water, which is raised about twenty-eight feet three times every two minutes; one bushel of coals, or two of coke, serving the boiler about two hours and three-quarters.

The *Atmospheric Steam-engine*, which is next in the order of invention, is now but little employed; indeed, if we except the mining districts where it is occasionally seen connected with the pumps for raising water, this species of engine is of very rare occurrence. The great merit of Newcomen's engine consisted, as we have already seen, in separating the parts in which the steam was to act, from those in which the water was to be raised; steam being employed merely for the purpose of displacing the air, and then forming a vacuum by condensation.

In this engine, steam of moderate expansive force is generated in the boiler *a*, *Plate* II. by the action of the fire in the furnace *b*.

c. The steam-pipe, through which the elastic vapour passes to the cylinder *d*.

d. The cylinder, furnished with a plug or piston, made to fit air-tight by means of a packing of hemp or any other elastic material.

e. The piston, connected with the working-beam *f* by means of a flexible chain and rod.

f. The working-beam, or lever, supported on the axis *g*.

h. The pump-rod, by the alternate elevation and depression of which the water is raised to any required height.

i. Injection-pipe, connected with the cold water cistern at *k*, and furnished with a small branch pipe *l*, to supply the upper side of the piston with water.

m. Eduction-pipe, furnished with a valve at *n*, to prevent the return of the water from the hot-well *o*.

p. Waste-pipe, to conduct the superfluous water from the top of the cylinder to the hot-well.

q. Injection and steam-cocks, alternately opened and shut by the plug-frame *r*, so that when the steam-pipe *c* is open to the cylinder, the connection with the injection cistern is closed, and *vice versa.*

s. The feeding-pipe, to supply the boiler with water, furnished with a cock at *t.*

v. The snifting-valve, by which, at every ascent of the piston, the air extricated from the condensing water is driven out by the pressure of the steam.

u. u. Two gauge-cocks, connected with pipes passing into the boiler, the one longer than the other, to ascertain the depth of water. Should one of these furnish steam and the other water, the latter may be considered at the required height. But if on the contrary both give steam, or both water, it is too high or too low.

w. Forcing-pump, worked by the main beam, for the supply of the injection-cistern, with which it communicates by means of the pipe *x x.*

When the atmospheric engine is set to work, the boiler must be filled rather more than half full of water, and the steam having attained a pressure of about one pound on each square inch of the boiler, the pump-rods will preponderate, and the piston be drawn to the top of the cylinder. In a few moments the elastic vapour will be seen to issue from the snifting-valve *v*, and the communication with the boiler being then closed, the injection water must be admitted, which condenses the steam, and of course forms a vacuum beneath the piston. The downward pressure of the atmosphere being now unbalanced by any resisting medium beneath, acts upon the piston with a force proportioned to its diameter, and it is made to descend with considerable velocity, at the same time raising the pump-rods *h w* connected with the opposite end of the beam.

In adjusting the working beam, it is necessary to allow the end connected with the pump-rods to preponderate, and this is accomplished by means of a moveable counter-weight. When an engine is erected on a mine, where the depth of the shaft is continually increasing, the quantity of water first lifted by the pumps being small, the engine must work slow, and the counter-weight be in proportion, allowance being made for the lightness of the pump-rods, which increase in weight with the progress of the mine. In the early stages, however, the injection must be very sparingly applied, so as to condense imperfectly within the cylinder, or the piston will descend with such velocity as to destroy the whole apparatus.

The boiler of Newcomen's engine was placed immediately beneath the cylinder; but this arrangement has, in the later engines, been materially improved; the boiler being now detached from the engine-room; and by this plan a considerable saving in the height of the engine-house is also effected.

We have already stated that Mr. Watt's great improvement consisted in condensing the steam in a separate vessel: the internal part of the cylinder being kept at the temperature of boiling water, so that the continued waste of steam, and consequently of fuel, that occurred by forming a vacuum beneath the piston in the atmospheric engine, was in this case avoided.

The nature of this improvement will be best understood by reference to the *Single-acting Engine,* that forms the subject of *Plate* III. in which *a* represents the boiler, enclosed in a casing of brick work.

b. The steam-pipe, connecting the cylinder *c* with the boiler.

c. The cylinder, firmly attached to the floor of the engine-room by the bolts *d d,* and having its upper end enclosed by the cap *e,* through which the piston-rod is made to work air-tight.

f. g. The beam, working on its axis or fulcrum at *h,* the socket in which the axis revolves resting on the floor and wall *i.*

j. The pump-rod, suspended at the end *g* of the working-beam.

k. The piston-rod, connected by the parallel motion at *f* with the working-beam *f g.*

m. The condensing cistern, containing the air-pump *n,* the condenser, and hot-well *o*: a continued supply of water being procured by the action of the cold water pump *p*, while the overflow is carried off by a waste pipe to the well *q*, so that nearly the whole of the external part of the apparatus is kept at the same temperature as the surrounding atmosphere.

r. and *s*. The upper and lower steam-valves.

t. The exhaustion-valve.

v. The plug-beam, furnished with pins to give motion to the levers acting on the valves *r s t.*

u. Pump, to raise water from the hot-well *o* to supply the boiler, which is effected by the pipe *w w*, the small cistern *x* being provided with a valve and lever furnished with a wire passing down to the boiler at *y*. The lower end of the wire is attached to a weight, which by its descent opens the steel-yard valve, and allows the admission of an additional supply of water when evaporation renders it necessary.

z. The man-hole, or aperture, formed in the top of the boiler, by opening the cap of which the necessary clearing and repairs are effected.

The single-acting engine (to which we have thus briefly referred) is merely employed to raise water; the steam acting above the piston, while a vacuum is formed beneath. A more minute description of the very compact double-acting engine of Messrs. Fenton, Murray, and Wood, will best serve to explain the internal mechanism and mode of working one of these gigantic machines.

Before, however, we proceed to an examination of its internal mechanism, it may be advisable to take a brief view of the general arrangement of its parts as exhibited in an engine of twenty-horse power erected at Leeds.

Plate IV. AAA. Foundation walls and masonry of the building on which the engine is erected.

B. The steam-cylinder, enclosed in a jacket or casing of cast iron, to exclude the atmospheric air from the cylinder, which is thus kept at the temperature of boiling water.

CC. The pipe which conducts the steam from the boiler to the valve-tube DD.

EE. The eduction-pipe leading to the condenser F.

G. The air-pump, which with the condenser F is immersed in the cold water cistern H.

I. The cold water pump, which supplies the cistern H by the pipe J.

K. The hot water pump, furnished with the piston-rod P, the upper part of which is connected with the working-beam at Q, so that at each elevation of the beam a quantity of hot water is furnished to supply the waste by evaporation.

L. The piston-rod, connected with the parallel motion MM.

NO. Two rods, attached to the opposite ends of the working-beam, and connected with the air and cold water pumps IG.

Q. The working-beam, supported by the cast-iron column R.

S. The connecting-rod, the lower part of which is attached to the crank T, while the other is elevated or depressed by the alternate motion of the working-beam.

U. A spur-wheel, attached to the crank-shaft, and working in a pinion V, by which it gives motion to the fly-wheel W.

XYZ. Three beveled wheels; the first of which is attached to the crank-shaft, and by the intermediate wheel and shaft gives motion to the third, which by a concentric roller moves the valves.

The parts of the engine we have thus briefly noticed, differ but little from the ordinary double-acting engine of Messrs. Boulton and Watt; and it will be necessary to refer to the enlarged scale on the following plate for a more accurate description of the working of the valves, &c.

Plate V. Fig. 1 and 2, represent sections of the steam-pipes, valves, and communicating rods.

C. The steam-pipe, furnished with a throttle-valve at *a*, to regulate the supply of steam to the engine. This is effected by the action of the lever *b*, and connecting-rod *c*, which communicates the action of the governor *g* to the valve *a*; while a rotatory motion is communicated to the axis of the governor, by means of a band passing from a pulley on the crank-shaft to a similar pulley on the axis of the governor.

e. e. Two bent levers passing through a slit in the middle of the spindle, and turning upon an axis at *f*. The upper part of the spindle is furnished with a socket *h*, which is allowed to ascend when the centrifugal force of the governor increases. Should, however, its motion decrease, the balls *j j* will descend, while the socket *h* will ascend, and with it the lever *l*.

c. A rod connecting the levers *l* and *b*, which by their joint action communicate the motion of the governor to the throttle-valve *a*, so that when the engine is at rest the balls *j j* will also be resting against the arms *k k*, the upper end of the levers *e e* will be brought nearer to each other, and the rod *c* being raised, the throttle-valve will be turned in a horizontal direction, thus allowing a large portion of steam to pass through the pipe C.

DD. A pipe connecting the top and bottom of the cylinder with the throttle-valve *a*.

E. The eduction-pipe, passing down to the condenser.

The valves *n o* have each a cylindrical tube or spindle passing through the stuffing-boxes *r* and *s*, to the upper end of which are screwed two other stuffing-boxes *t* and *u*, so that both valves are allowed to slide up or down without permitting the steam to pass.

p. *q*. Two other valves similar to *n o*, whose spindles pass through the stuffing-boxes *t u*.

Fig. 2. Is a front view of the two sliding bars which are intended to give motion to the valves *n o p q*. These bars are kept in a perpendicular direction by the pieces *z z*, and the guide 1. In the lower ends of the bars are two friction rollers 3 3, which are acted upon by the two eccentric wheels 4 4, and raised and depressed alternately by the upward and downward stroke of the engine.

The horizontal shaft Z derives its motion from a similar shaft Y placed at right angles, communicating by means of beveled wheels with the crank-shaft.

9, 10, 11, and 12, are four arms, fixed to the bars *v v*, and *w w*, for the purpose of moving the valves.

13. A lever or handle revolving upon a stud screwed in the pipe E, which, by its action, is made to open and shut the steam-valves when the engine is first set to work.

18. A mercurial or barometer gauge for measuring the pressure of the steam above or below that of the atmosphere. One end of the barometer-gauge enters the steam-pipe DD, while the other is open to the atmosphere and furnished with a gauge.

The communication between the barometer-gauge and steam-pipe may be closed at pleasure by the stop-cock 19. In the lower or bent part of the tube is placed a quantity of mercury, and it will be evident that upon opening the cock 19, the pressure of the steam endeavouring to pass by the pipe DD, will be counterbalanced by the pressure of the atmosphere. Should, however, the elasticity of the steam exceed that of the atmosphere, the mercury will be raised in the outer leg of the gauge, and the difference in the altitude of the two columns will show the working power of the steam. When the altitude of the column 18 exceeds that of 19 two inches, the pressure of the steam will exceed that of the atmosphere nearly one pound per square inch.

A nearly similar instrument is also employed to ascertain the degree of rarefaction in the condenser. It consists of a bent iron tube 21, the lower end of which opens to the condenser. The mercury is poured into the tube at the open end 23, and the stop-cock 22 being opened, the mercurial column at 23 will be depressed, while that on the opposite side will be raised in a proportionate degree. This effect is produced by the vacuum formed in the condenser. If the condenser and air-pump are in good order, the mercury will descend about fourteen or fifteen inches, which will indicate a pressure of so many pounds upon the square inch. So that if we refer to the two gauges, it will be found that the total amount of power, or acting force upon the piston, will be represented by the difference in the altitude of the two mercurial columns added together.

To put the engine in action, the fly-wheel W, *Plate* IV. must be turned till the crank T is in a horizontal direction, when the piston will be in the middle of the cylinder B, and the eccentric wheel 4 on the upper side of the shaft Z. The bar *w w*, will also be raised, together with the valves *o* and *p*, and the handle 13 being turned, a passage will be opened for the steam to blow from the pipe C, and thus expel the atmospheric air, which previously filled the different parts of the engine. When this is effected, and the temperature of the engine raised, the lever 13 must be turned to its original situation; the bar *v v*, together with its valves, will descend, and the steam will be shut off from the upper side of the cylinder; while, at the same time, the passage will be stopped between the under side of the piston and the condenser. The injection-cock must now be opened, which will admit a small jet of cold water into the condenser, and a vacuum will be formed above the piston, while the steam is entering beneath with a pressure equal to or greater than that of the atmosphere.

The piston-rod being thus made to ascend in the cylinder, the opposite end of the beam Q will be depressed in a proportionate degree, and the rod S, as well as the crank T, will also descend, and a rotatory motion be produced. The fly-wheel will also have acquired a sufficient degree of momentum to carry the crank past the perpendicular, and the piston will have arrived at the top of the cylinder; the situation of the valves being reversed by the action of the excentric wheels, and a continuous motion is thus produced.

The *High-pressure Engine*, in its most simple form, may easily be described by reference to the following diagram.

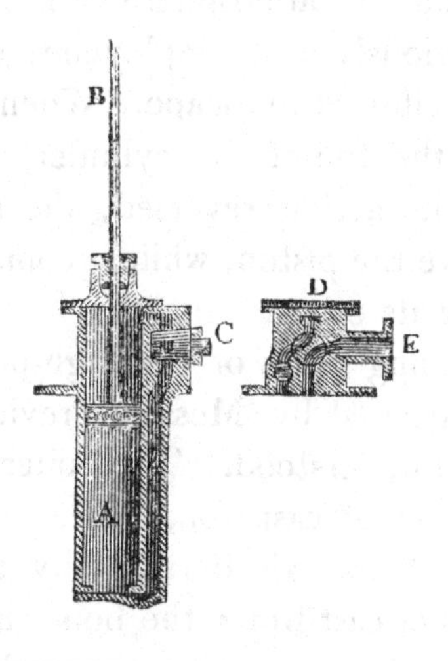

The cylinder A is furnished with a piston and rod B, the latter being made to fit air-tight in a stuffing-box at the top of the cylinder. A four-way cock C is also provided for the admission of highly elastic vapour, and its subsequent discharge into the atmosphere. The action of the four-way cock will be best understood by the section D; in which E represents the waste-pipe connected with the chimney, while two other apertures serve to convey the steam alternately to the upper and under side of the piston, and a third communicates with the steam-boiler. So that if we suppose the piston to be in an ascending direction, and the steam of course entering the cylinder beneath, a communication will at the same time be formed between the upper side of the piston and the atmosphere, while the steam that had previously been employed to depress the piston is now allowed to escape. When the piston has reached the top of the cylinder, the cock is turned, and its action reversed, the steam now entering above the piston, while a communication is formed for its escape beneath.

The remaining parts of the high-pressure engine, as constructed by Messrs. Trevithick, may very easily be understood. The boiler consists of a large cylinder of cast iron, made very strong, and placed with its axis horizontally upon short feet or pillars of cast iron; the boiler has a flanch at one of its ends to screw on the end or cover, which has the requisite openings for the fire-door, the man-hole, the exit for the smoke, and the gauge-cocks. The fire is contained within the boiler in a cylindrical tube of wrought iron, which is surrounded with water on all sides; one end of this tube is attached to the end or cover of the boiler, and is divided into two parts by having the fire-grate extended across it; the fire-door closes the opening in the upper half, which is the fire-place, the lower half forming the ash-pit; the tube extends nearly to the end of the boiler, where it is reduced in size, then doubles, and returns back in a direction parallel to the first tube or fire-place, to form the flue, till it arrives at the end of the boiler, through which it passes at the side of the fire-door, and it is then conducted from it into the chimney, thus carrying off the smoke.

At the part where the flue enters the chimney, is a small door to remove any soot that may have accumulated. On the top of the boiler is a safety-valve, kept down by a lever and weight, to allow the steam to escape in case it becomes so strong as to endanger the bursting of the boiler. The cylinder of the engine stands in a perpendicular direction, and is enclosed within the boiler, except a few inches of its upper end, at which the four-passaged cock is situate, and the flanch which screws on the lid, with the stuffing-box for the piston-rod to pass through. The boiler has a projecting neck, into which the cylinder is received, and it is fastened in its place by a flanch round the upper end of the neck of the boiler, which is united by screws to a similar flanch projecting from the cylinder. The upper end of the piston-rod is fastened to the middle of a cross-bar, which is placed in a direction at right angles to the length of the boiler, and guided in its ascending and descending motion, by sliding between two perpendicular iron rods, fixed to the boiler, parallel to each other, being connected together at top, and firmly supported there by two diagonal stays extending from the other end of the boiler, and secured to the flanch, which screws on the end of the boiler. At the ends of the cross bar of the piston-rod, the two connecting-rods are jointed, and the lower ends of them are connected with two cranks, fixed upon an axis, extending across beneath the boiler, and under the centre of the cylinder; the axis is supported in bearings made in the legs which support the boiler, and the fly-wheel is fixed in it. One of the cranks is formed by a pin, which is fixed into the arm of the fly-wheel at the same radius as the opposite crank. The fly-wheel is situate close to the side of the boiler, and the pin for the other crank is fixed into the arm of a large cog-wheel, fixed on the axis of the fly-wheel at the opposite side of the boiler. This cog-wheel communicates the power of the engine to other cog-wheels. As the piston is alternately forced up and down by the pressure of the steam, it carries the cross-bar with it, and by the connecting-rod turns the two cranks, together with the fly-wheel, and other connecting machinery.

The *Double-cylinder Expansion Engine*, constructed by Mr. Woolf, possesses considerable advantages over that invented by Mr. Hornblower, with whom, it appears, the idea of constructing an engine with two cylinders originated: the only novelty, however, in Mr. Woolf's engine, consists in the application of high-pressure steam, and in the proportioning of the cylinders to the elasticity of the vapour employed. Having already explained the principle upon which Mr. Woolf's engine is constructed, it may be enough for our present purpose to describe its action, with reference to the annexed diagram.

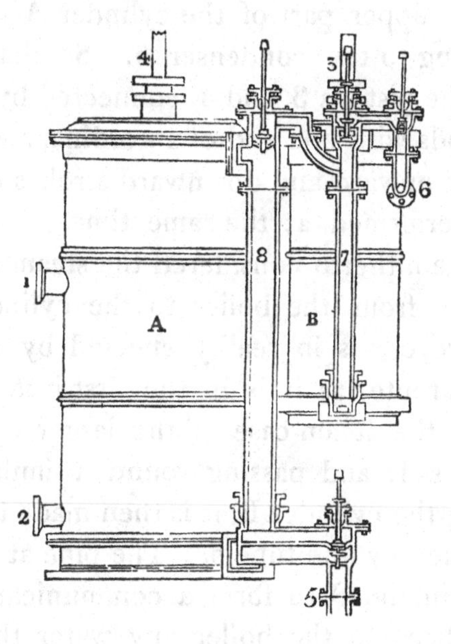

A and B represent the two cylinders, in the larger of which the steam is allowed to expand itself, after passing from B. The steam, which in the first instance is of considerable elasticity, is admitted to the cylinder B, by the tube and valve 6, and entering the cylinder above the piston 3, impels it to the bottom. When this is effected, a communication is opened between the upper part of the cylinder B, and the under side of the piston 4, in the larger cylinder. The communication between the cylinder B and the steam-pipe 6, is now reversed, and the steam is made to press on the under side of the piston 3; a communication being at the same time formed between the upper part of the cylinder A, and the pipe leading to the condenser 5. So that if we suppose the pistons 3 and 4 connected by means of their rods with one end of an ordinary working beam, the upward and downward strokes of each will be performed at the same time.

We have hitherto considered the steam as passing direct from the boiler to the cylinder B; this, however, is in reality effected by a more circuitous route, as it is in the first instance admitted to the steam-case of the larger cylinder, by the pipe 1, and passing round a similar case, encircling the cylinder B, it is then made to enter the cylinder by the tube 6. The pipe at Fig. 2. is merely intended to form a communication for carrying back to the boiler any water that may be produced by condensation in the steam-case, before the engine arrives at a proper temperature for working.

Having described the nature of Mr. Woolf's engine, it may now be advisable to examine the boiler, by which he proposed to generate steam of sufficient elasticity for the use of the small cylinder, which requires vapour of great expansive force.

The boiler represented by the diagram beneath, consists of a series of tubes of cast iron, connected by screw-bolts with the under side of a larger vessel, or magazine of the same material. This is furnished with four, and in some cases with five apertures; the first of which, A, is intended for the admission of water, to supply the waste which continually arises from evaporation, which is effected by means of a small forcing-pump, as it will be evident that the column must be carried to a considerable height, before its weight can so far overcome the resistance of the steam within the tube, to allow of its entering by the ordinary method.

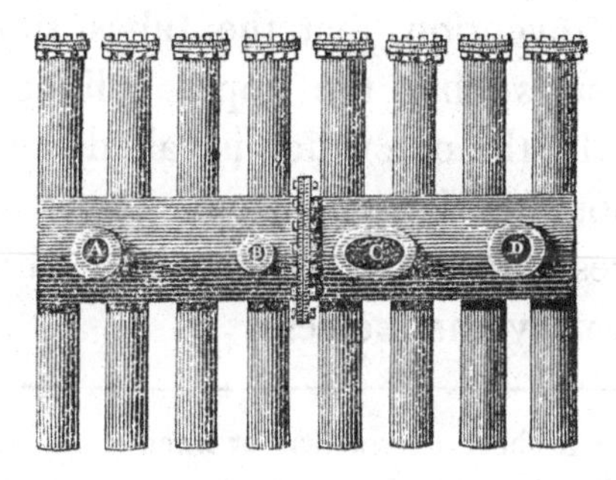

Mr. Woolf usually employs two safety-valves which are placed at B, while C represents the man-hole, and D the pipe by which the steam is conveyed to the engine.*

* These, as well as the flanches, and other steam-tight fastenings of a permanent nature, are usually connected by screw-

In Mr. Woolf's specification, a method is pointed out for applying this plan to the boilers of steam engines already in use, by placing a series of cylinders beneath the present boilers, and connecting them with each other, and with the boiler above. The tubes may be made of any kind of metal, but cast iron is the most convenient; their size may also be varied, but in every case care should be taken not to make the diameter too large; for it must be remembered, that the larger the diameter of any single tube is in such a boiler, the stronger it must be made in proportion, to enable it to bear the same expansive force of steam as the smaller cylinders. It is not essential, however, to the invention, that the tubes should be of different sizes; but the upper cylinders, and more especially the one which is called the steam-cylinder, should be larger than the lower ones, it being the reservoir, as it were, into which the lower ones empty themselves.

---

bolts and nuts; a sheet of woollen or linen cloth coated with cement, being first introduced to unite the intervening surfaces. The cement best adapted for this purpose, from its durability and power of withstanding the action of steam, may be thus prepared:—Take two ounces of sal ammoniac, one ounce of flour of sulphur, and sixteen ounces of cast-iron filings; these, after being well mixed in a mortar, must be placed in a dry situation, and when wanted for use one part of the above mixture must be blended with twenty parts of clean filings, and saturated with a little water. On being applied to the joint it will shortly become as hard as the metallic surface on which it is placed.

The following general directions are given respecting the quantity of water to be kept in a boiler of this construction; viz. it ought always to fill, not only the whole of the lower tubes, but also the great steam-cylinder, A, to about half its diameter, that is, as high as the fire is allowed to reach. And in no case should it be allowed to get so low as not to keep the vertical necks, or branches which join the smaller cylinders to the great cylinder, full of water, for the fire is only beneficially employed when applied, through the medium of the interposed metal, to water, to convert it into steam; and indeed, the purpose of the boiler would in some measure be defeated, if any of the parts of the tubes thus exposed to the direct action of the fire, should present a surface of steam instead of water, to receive the transmitted heat; this, however, must, more or less, be the case whenever the lower tubes, and even a part of the upper, are not kept filled with the water.

Respecting the furnace for this kind of boiler, it should always be so built as to give a long and waving course to the flame and heated air, so that they may, in the most effectual manner, strike against the sides of the tubes which compose the boiler, and so give out the greatest possible portion of their heat before they reach the chimney: unless this be attended to, there will be a much greater waste of fuel than necessary, and the heat communicated to the contents of the boiler will be less from a given quantity of fuel.

Mr. Maudslay, in his *Portable Engine*, dispenses with the beam usually employed for connecting the fly-wheel crank with the piston-rod; and, in this respect, as well as in the working of the valves, his engine materially differs from those we have already described.

Plate V. Fig. 1. Front elevation of a ten-horse power engine.

Fig. 2. Longitudinal section of ditto, on the centre line.

Fig. 3. End view of ditto.

A. Cast-iron frame of the engine.

B. The cylinder.

C. The piston, furnished with a rod D, and a cross head and socket E.

F. Guide wheels, which keep the piston and rod in a vertical position.

G. Frame for ditto, in which the wheels FF are made to work.

H. Side rods, which serve to connect the cross head E with the double crank II.

II. Two cranks, made to turn in the plummer-block, or bearing, JJ, at each side of the frame, and to which the fly-wheel shaft K is connected by a coupling-box or clutch, at the end next the engine.

K. Fly-wheel shaft, working in a plummer-block on the wall.

L. Coupling-box, connecting the engine with the fly-wheel shaft.

M. The fly-wheel.

NN. Two excentric wheels, supported by the crank-shaft K, the action of which give motion to the two beams O and T, by means of the connecting-rods PP.

O. The beam which works the cold-water pump S.

PP. Two connecting-rods.

Q. The double bearing, on which the cold-water pump-beam works.

R. A rod which serves to connect the bucket of the cold-water pump with the beam O.

S. The barrel of the cold-water pump.

T. Beam which works the air and hot-water pumps, and to which motion is communicated by the connecting-rods P, as before described.

U. The slings which connect the air-pump rod to the end of the beam T.

V. The double bearing, or centre, on which the air-pump beam T works.

W. The air-pump bucket.

X. Air-pump cylinder.

Y. Hot-water pump, worked by a small rod, attached to the air-pump beam.

Z. Feed-pipe, to supply the boiler with hot water.

a. Cross rail on which a guide is fixed to confine the air-pump rod in a vertical position.

b. The condenser.

c. The cold water cisterns, connected by a pipe *d.*

e. Eduction pipe, or passage for the steam from the cylinder to the condenser.

f. Injection cock, to admit the cold water into the condenser.

g. Foot valve, at the bottom of the air-pump, and communicating from thence to the condenser.

h. Hand gear, for stopping or starting the engine.

i. A rod, connecting the hand gear with an excentric *k*, fixed on the crank-shaft ; the action of which communicates a vibratory motion to the rod *i*.

l. Connecting-rod, and double-ended lever *m*, fixed at the extreme end of a spindle, while a beviled wheel is attached to the other ; the latter of which works the spindle of the steam-cone *n*.

o. The steam-cone, or cock, for admitting the steam from the boiler to the cylinder ; beyond which is a contrivance for shutting off the steam, at the half, or any other part of the stroke, by which a very considerable saving in the steam, and consequently in the fuel, is effected.

It will very readily be seen, that the cone employed in this engine, for regulating the passage of the steam from the boiler to the cylinder, differs very materially from the valves in Messrs. Murray and Wood's engine, and a slight examination of the sectional view in *Plate* V. will shew that the greater degree of friction that must of necessity attend the former contrivance, is more than compensated by its superior tightness and simplicity.

A. represents an end view of the cylinder and steam-cone.

B. Side view of ditto.

C. Plan of ditto, taken at the horizontal line D.

E. Steam-pipe.

F. Pipe, communicating with the condenser.

G. Steam-cone, ground into its seat, and connected with the grease-cup H, by the means of which a regular supply of oil is furnished.

I. Plan of the steam-cone and passages, by which a communication is alternately opened between the steam-pipe and the upper end of the cylinder, and the condenser and the lower end of the cylinder, and *vice versa.*

When we consider the reciprocating steam engine in its present most improved state, both with respect to the admirable expedients for adapting it to the end proposed, and the skill displayed in the workmanship, we may almost venture to conclude that it has reached its utmost degree of perfection; and yet it must be acknowledged, that it absorbs nearly half the power of the steam employed in friction, and in alternating its movements. This fact will be apparent by calculating what pressure on the piston of a reciprocating condensing engine, would be required to produce its nominal power; and it will be generally found, that (with the common speed) this will be obtained by accounting only from six to seven pounds pressure, per square inch, on the piston of small power engines; seven to eight pounds as to engines of from ten to thirty-horse powers; and from eight to nine pounds as to engines of larger powers, though the actual pressure on the piston is nearly sixteen pounds.

We have seen, that steam engines were, in the first instance, used for raising water, for which purpose the alternating motion of the beam is well adapted; at present, however, by far the greater number of reciprocating steam engines are required to impart a rotatory motion to the machinery attached to them.

The loss of power, to which we have already alluded, together with the expense of the construction of the reciprocating engine, have induced numerous attempts to invent an engine imparting a rotatory motion in the first instance; and the recent application of this prime mover to the purposes of navigation, has also acted as an additional stimulus to the attainment of so desirable an object, the inconveniences of a reciprocating engine being most sensibly felt in steam-vessels. Hitherto, however, those attempts have been attended with only partial success; for though many patent inventions have come under our observation, the principal of which will be found in the Appendix attached to this work, they have altogether failed in attaining any decided superiority over the reciprocating engine, either from excessive friction, or the expense and nicety of workmanship required both in their construction and repairs.

Those difficulties appear to be obviated in Messrs. Masterman's *Rotatory Engine*, in a much greater degree than in any other that has come to our knowledge : the entire friction of one of those engines (without a condenser) having been proved, from actual experiment, not to exceed half a pound per square inch on the valves; the expense of construction being very considerably less than that of reciprocating engines, particularly in the larger powers, and the extraordinary simplicity of its parts securing it from almost any expense for repairs.

For steam navigation it appears admirably adapted, and when used with mercury instead of water, combines, in an eminent degree, economy of space and fuel; and this, in the latter case, it will be evident must be very considerable, on account of the almost total absence of friction.

Plate VII. Fig. 1. is a vertical and central section of the revolving part of the engine, called the troke, which is composed of a centre *a*, called the nucleus, of six hollow arms *b*, 1 to 6, called radii, and of a hollow ring *cc*, called the annulus.

Fig. 3. represents the nucleus; one end *m* is a perfect circular plane, called the face; six holes of similar figure and dimensions are sunk in the face at equal distances from each other, following a direction parallel with the axis *e*, until half way through the nucleus, then, assuming a direction at right angles with the axis, they open in the periphery of the nucleus.

The axis passes through the centre of the nucleus at right angles to the plane of the face. The annulus consists of six equal parts, in each of which is fixed a steam-tight valve, exactly similar, and opening in the same direction by a hinge placed in the side of the annulus nearest the axis.

The rods which form the hinges of the valves, project through stuffing-boxes in the side of the annulus; and on each of these projections is placed a lever, at such an angle with the valve as to point to the axis when the valve is half open; and at the extremity of each lever is a weight *d*, more than sufficient to counterpoise the weight of the valve against which it acts. In the side of the annulus nearest the axis, are six holes at equal distances from each other; these holes are connected with the holes opening in the periphery of the nucleus by means of the hollow radii *b*, 1 to 6, thus forming a steam-tight communication between each hole in the face and the inside of the annulus.

Fig. 4. is a section of a metal plate or mask, which is of equal diameter with the face, having one side ground perfectly flat.

Through the centre of the mask is a circular

hole *e* to admit the end of the axis. In the space surrounding this hole there are three other holes, *p*, *i*, and *l*. The holes *i* and *l* are each of such dimensions as to extend over one of the holes in the face and the adjoining space; and the space in the mask between the holes *i* and *l* is of such dimensions, as just to cover completely one of the holes in the face.

The holes *p i* and *l* terminate in lateral apertures, as represented in Fig. 5. The hole *l* is connected with the boiler by a steam-pipe; the hole *i* with the condenser, if one be used, or discharges the steam into the air. To the hole *p* is fixed a perpendicular pipe, rising above the level of the troke.

The mask does not revolve, but is kept closely pressed against the face by means of the nuts and rings, Fig. 6. which are fixed on the end of the axis *e*; and it is maintained in such a position, that the space between *l* and *i* is just above the level of the centre of the troke, and on the side of the axis nearest the closed valve in Fig. 1.

Fig. 2. is a section on the plane of the axis of the troke mounted on its axis, together with the mask applied to the face, and of a reservoir *k* at the top of the pipe *h*, for supplying the interior of the annulus with water. To the end of the axis farthest from the face, the machinery to which it is proposed to impart motion is affixed.

The engine is worked by steam and water as

follows. The annulus is in the first place half filled with water, either admitted cold, and heated by suffering the steam to flow into it through the pipe *e*, or admitted in a heated state from the boiler. On the steam valve being opened, the steam enters the hole *l* in the mask, Fig. 4. through the pipe *l*, Fig. 5. and passes thence through the hole in the face which happens to be opposite to the aperture in the mask, and enters the annulus; then rising through the water, it is stopped by the valve *d*, immediately above the radius by which it entered, which will then be closed. The steam resisted by the valve, acts against the surface of the water below it, and pressing it downwards, proportionably raises it on the opposite side of the annulus, until the pressure of the column of water acting against the closed valve, through the medium of the steam, is sufficient to overcome the resistance. The troke is now made to revolve, and, as it revolves, each of the holes in the face communicates in succession with the hole *l*, and, by this construction, one entire hole in the face, or parts of two equal in proportion to one, is always in communication with the hole *l*; so that there is a continual flow of steam into the annulus, causing the water, through its medium, to exert a constant and uniform pressure on the valves as they ascend. The holes in the face, as they pass in succession from the hole *l* to the hole *i* are entirely closed by the space between them; and, immediately on communicating with the hole *i*, the steam rushes from the annulus through that hole into the condenser, or into the air; and the pressure of steam being thus removed from the valves, they will open by the gravity of the weights *d*, as they begin to descend, and thus allow the column of water to remain on that side of the annulus. Thus a uniform rotatory motion is produced and maintained as long as the steam continues to flow into the annulus, and acting with a force proportionate to the difference of level in the water.

In Fig. 1. the troke is represented as revolving, and the steam flowing into the annulus, through the radius *b* 1; *f* represents the steam in the annulus between the closed valve and the depressed surface of the water; *g* the water raised on the opposite side of the annulus, while the remaining or darkest part of the annulus is where the valve and the upper surface of the water are relieved from pressure, the steam having discharged itself through *b* 6.

The steam may be admitted through a radius more or less horizontal, according as the column of water is higher or lower, by means of an inner mask, which changes its position; the closing spot of the valves may also be regulated accordingly, by means of catches acting on their levers. The troke is of cast iron, and, to prevent condensation, it is enclosed in a steam-tight case.

From this brief examination of Messrs. Masterman's engine, it will, we think, be apparent, that the troke alone performs the united function of cylinder, piston, beam, crank, and fly-wheel; thus ensuring a decided superiority over the reciprocating engine.

The advantages resulting from the use of steam engines have, in some cases, been considered as fully equipoised by the smoke and noxious effluvia which proceed from their capacious vomitories; and this, in large manufacturing towns, is indeed an evil of some importance, to obviate which a variety of contrivances have been suggested.

The first attempt at consuming smoke, appears to have been made by M. Dalesme, a French engineer, who exhibited a contrivance of this description at the Fair of St. Germain's in 1685.* In 1785 Mr. Watt obtained a patent for the construction of an economical furnace, which not only consumed the smoke, but employed it as an useful auxiliary in increasing the heat. To understand this it will be necessary to observe, that the dense smoke which is usually discharged at the top of the chimney, is in fact, so much good fuel, which requires but a sufficient supply of oxygen to render it fit for combustion.

Mr. Watt accomplished this in his early engines by stopping up every avenue to the chimney, except such as might be left in the interstices of the ignited fuel, and the smoke from the fresh coal was

* Vide, Transactions of the Royal Society, vol. xvi. p. 78.

consumed by passing through the burning fuel or coke.

A variety of improvements have since been introduced, an account of which will be found in Appendix (B,) and we shall content ourselves, in the present instance, with briefly noticing those that appear best calculated to answer their intended purpose.

Mr. Roberton's plan is nearly similar to that employed by Mr. Watt. The opening through which the fuel is introduced into the furnace is shaped like a hopper, from the mouth or entrance of which it inclines downward to the place where the fire rests on the bottom grate. The fresh coals contained in the hopper answer the purpose of a fire-door, and the principal point to be attended to in the management of this furnace is, that the hopper shall be so filled with small coal as to prevent as much as possible the passage of atmospheric air by the hopper. Beneath the lower part of the hopper the furnace is provided with front bars, which serve to admit air among the fuel which is undergoing the process of coking in the lower part of the hopper, and at the same time offers a ready mode of forcing the ignited fuel thus deprived of its smoke upon the centre of the burning mass, where it is completely consumed, while an additional supply of fresh coal falls down the hopper to supply its place. By this arrangement, and the judicious admission of a thin stratum of fresh air, by a valve placed near the mouth of the hopper, the quantity of smoke is considerably reduced, the whole of the fuel being brought into a state of ignition before it has arrived at the middle of the burning mass, and a sufficient supply of oxygen is furnished by the air-valve for the purpose of combustion.

Sir William Congreve's invention consists in the application of chalk, or any other calcareous substance convertible into lime by means of heat, as auxiliaries to the ordinary articles of fuel. This is effected by converting the furnace into a species of lime-kiln, in which the mass of heated coal is employed not only to heat the boiler, but calcine a large quantity of the above substance; thus concentrating and keeping in action a most powerful heat in aid of the ordinary operation of the fuel.

The following is the substance of a series of experiments and calculations, made in the Royal Laboratory at Woolwich, which serve to shew the great advantages attendant on the adoption of this method.

Thirty gallons of water were evaporated in seven hours by half a bushel of coal, weighing forty-two pounds, calcining at the same time one bushel and a half of lime. Thirty-four gallons of water were afterwards evaporated in the same time, without burning the lime, and required one bushel and a half, or 126 pounds of coal. These experiments were afterwards repeated, and the same results obtained.

It appears therefore, from these trials, that half a bushel of coal, with lime, generates very nearly the same quantity of steam as one bushel and a half without the lime. This however may be better illustrated by the following statement of the comparative expense.

		s.	d.
First experiment.—Half a bushel of coal		0	7
	One bushel and a half of chalk ..	0	2
		0	9
Second experiment.—One bushel and a half of coal ..		1	9

In the first experiment the lime produced by this species of burning may fairly be averaged at nine pence, so that, compared with the present mode, the saving on evaporating thirty gallons of water by means of the chalk, where a ready mode of disposing of the lime can be devised, is nearly 1*s.* 9*d.* or the total expense of the fuel.

Mr. Parkes employs an air-valve, somewhat similar to that of Mr. Roberton, though placed in a different part of the furnace; and either of these plans, if properly managed by the fire-man, would fully answer the end for which they were intended; but unfortunately this requires a degree of mechanical skill and attention seldom found in this class of persons; and though the nuisance may be abated for a short time, or while the engine is under the immediate superintendance of the engineer, no very permanent benefit has yet been found to accrue. To remedy this, Mr. Brunton proposes to employ a mechanical apparatus completely beyond the control of the attendant, whose attention may in this case be almost entirely dispensed with. In Mr. Brunton's furnace the grate bars are made to revolve in an horizontal direction beneath the boiler, by which means the heat is uniformly distributed over the whole of its lower surface, and a regular supply of coal is furnished from a hopper above.

To effect this the axis upon which the grate turns is connected with the steam engine itself; and for a boiler of five feet diameter, it is made to perform about one revolution per minute. Every time it arrives at a certain point, the channel from the coal hopper is opened; and in order to prevent the air from passing down through the coal, the patentee in his specification describes a rim, upon which the regulator is intended to lay, descending into a trough for the purpose of forming a water or sand valve. There is also a regulator to the feeder, connected with the damper, so that if the boiler become too hot, or the pressure of the steam increase, the quantity of coal supplied should be diminished in a proportionate degree. The nature of this very ingenious apparatus will however be more fully understood by a reference to the improved furnace, &c. erected at Messrs. Smith and Liptrap's distillery, Whitechapel.

Plate VIII. AA. Waggon boilers, to which the supplementary boilers BB are attached; the smaller or supplementary boilers being placed immediately over the fire, while the larger boilers derive an additional supply of heat from the passage of the chimney C.

D. Chimney doors.

EEE. Hoppers by which the coal boxes FFF are supplied with fuel.

FFF. Coal boxes furnished with sliding plates, through the openings of which the coals are allowed to fall on the ignited fuel.

GGGG. Steam-pipes joining the waggon and supplementary boilers.

H. Furnace door attached to the supplementary boiler by a cement joint.

II. Doors opening into the air-flues, to assist in the combustion of the smoke, and to withdraw the dust that may fall over the edge of the fire-grate.

K. Axis or spindle upon which the grate is made to revolve; the motion being communicated direct from the engine by the pinion and wheels L.

M. Foundation plate, in which are formed the pivot holes for the axis K and the upright shaft N.

O. Feed-pipes of the waggon boilers.

P. Steam-pipe leading to the engine.

Q. Pipe communicating with the safety-valve V

R. Horizontal shaft communicating with the vertical axis N, and also with the engine by which the whole apparatus is turned.

SS. Chains attached to the damper chains, by which the lever T is moved, and the wedge U made to rise or fall with the damper plate; so that when the steam is in excess, it may diminish the supply of coals in proportion to that excess, and *vice versa.*

V. Safety-valve.

W. Self-acting stop-valve, to prevent the steam passing from one boiler to the other when two boilers are used.

X. A rod connected with the lever Y, which by pressing upon the stop-valve, closes the communication between the two boilers, when a reduced supply of steam is required.

Z. Gauge pipes to ascertain the amount of water in the boiler.

a. Man-hole of waggon boiler, furnished with an internal safety-valve *b.*

c. Stone float within the boiler.

d. Bridge walls.

f. Sand trough, in which revolves a thin plate attached to the fire grate, to prevent the air passing in any other way than through the bars.

h. The fire bricks surrounding the grate bars.

i. A scraper attached to the grate, and, which revolving with it, cleans the air-flue.

From the above description it will be evident, that the great advantage arising from the employment of this apparatus, consists in an equable supply of coal, and in the smoke arising from its combustion having to pass over the entire mass of burning fuel prior to entering the chimney. By these means, the greater part if not the whole of the smoke is consumed; and it will be evident that the inflammable materials of which it is composed, will furnish an additional supply of valuable fuel, which would otherwise be thrown unconsumed into the atmosphere. As, however, direct experiment is the only sure test, by which this, or indeed any project, can fairly be tried, we annex the results arising from two experiments; the one made at the distillery of Messrs. Liptrap and Smith, Whitechapel, London, to whose kindness we are indebted for the means of completing the above description, and the other at the Old Union Mill, Birmingham.

At the Old Union Mill, Nine days experiment,

Common Furnace consumed	-	465 cwt.
Fire Regulator	- - -	290

The Whitechapel Distillery, Eighteen Days experiment,

Common Furnace consumed	-	284 bushels.
Fire Regulator	- - -	194

APPENDIX (A).

*List of Patents for the Steam Engine, with an Analytical Account of those more immediately connected with its Improvement and general Application to the useful Arts.**

T. Savery, London, July 25, 1698.

This patent, which is the first upon record, for an invention in which steam was employed as a prime mover or principal agent in hydraulics, describes two modes of effecting this very desirable object. In the first, it is used merely to produce a vacuum by condensation; and, in the latter, the impellent or expansive force of the steam is made to act upon the surface of the fluid to be raised, and by its pressure in a close vessel, the water is driven up a connecting tube to the required height.

In this engine a vacuum being first formed by the condensation of steam, the water was afterwards raised by the pres-

* A complete list of the patent-right inventions connected with this branch of our manufactures, has long been a desideratum; while a reference to the chronological arrangement will shew the progressive improvements that have been effected in its construction. In addition to this, the future experimentalist may derive considerable benefit from the labours of his precursors thus at one view presented to his notice. It is scarcely necessary to add, that a large portion of these exclusive monopolies are of little value beyond that of swelling the fees of the patent office; many of them being precisely the same in both principle and application.

sure of the atmosphere to a given height from the well into the engine, and then forced out of the engine up the remaining height by the pressure of steam upon its surface. This action was performed alternately in two receivers; so that while the vacuum formed in one was drawing up from the well, the pressure of steam in the other was forcing up water to an elevated reservoir, and by this means a continued stream of water was produced.

T. Newcomen and J. Cawley, 1705.

This engine has been very fully described in a preceding page; to that, therefore, with a reference to the plate, we shall beg to refer the reader.

J. Hull, London, Dec. 21, 1736.

The mode of propelling vessels by the application of paddle-wheels now so generally adopted, appears to have been originally suggested by this patent. Mr. Hull proposed to employ the atmospheric engine of Newcomen, which, by means of a crank communicating with the working beam, imparted a rotatory action to the wheels or paddles which were placed at the bow of the vessel.

James Brindley, Lancashire, 1759.

The boiler in this engine was proposed to be made of wood and stone, with a cast iron fire-place within side of it, and surrounded on all sides, so as to give its heat to the water. The chimney was an iron pipe or tube, also immersed in the water of the boiler; and this plan Mr. Brindley expected would save a considerable portion of the fuel.

Blakey, 1766.

This patent which consisted in an improvement upon Savery's engine, was in principle similar to that of Dr. Papin.

In this case a quantity of oil was placed in the receiver, which, rising to the surface, formed a species of piston or float between the surface of the water and the hot steam; thus preventing the continued condensation of elastic vapour, which would necessarily occur in engines upon the original construction.

To effect this two receivers were to be used, one in the same situation as Savery's, which was to receive the air; and the hot steam, when admitted into it, forced the air to descend by a pipe to the second receiver, which was at the bottom of the well from whence the water was expelled, and proportionably raised in the force pipe.

J. Watt, Birmingham, Jan. 5, 1769.

To the great and comprehensive genius of the late Mr. Watt, and the spirit of rivalry which was excited in the mechanical world on the publication of this patent, which, though the earliest, is certainly the most important of his inventions, may be ascribed the completion of those improvements that have subsequently been effected in the steam engine. This patent, the term of which was prolonged for twenty-one years from the expiration of the original grant, contains the following principles, which, for their importance, we insert in the author's own words: "First, That the vessel in which the powers of steam are to be employed to work the engine, which is called the cylinder in common fire engines, and which I call the steam vessel, must, during the whole time the engine is at work, be kept as hot as the steam that enters it; first, by enclosing it in a case of wood, or any other material that transmits heat slowly; secondly, by surrounding it with steam or other heated bodies; and, thirdly, by suffering neither water nor any other substance colder than steam, to enter or touch it during that time.

"Secondly, In engines that are to be worked wholly or

partly by condensation of steam, the steam is to be condensed in vessels distinct from the steam vessels, although occasionally communicating with them. These vessels I call condensers; and while the engines are working, these condensers ought to be kept as cold as the air in the neighbourhood of the engines, by the application of water, or other cold bodies.

" Thirdly, Whatever air, or other elastic vapour, is not condensed by the cold of the condenser, and may impede the working of the engine, is to be drawn out of the steam vessels by means of pumps connected with the engine.

" Fourthly, I intend in many cases to employ the expansive force of steam to press on the pistons, or whatever may be used instead of them, in the same manner as the pressure of the atmosphere is now employed in common fire engines. In cases where cold water cannot be had in plenty, the engines may be wrought by force of steam only, by discharging the steam into open air after it has done its office.*

" Fifthly, Where motions round an axis are required, I make the steam vessels in the form of hollow rings, or circular channels, with proper inlets and outlets for the steam, mounted on horizontal axles, like the wheels of water-mills. Within them are placed a number of valves, which suffer bodies to go round the channels in one direction only. In these steam vessels are placed weights so fitted to them as entirely to fill up a part or portion of their channels, yet rendered incapable of moving freely in them by the means hereinafter mentioned or specified. When the steam is admitted into these engines, between the weights and the valves, it acts equally on both, so as to raise the weights to

* This should not be understood to extend to any engine where the water to be raised enters the steam vessel itself, or any vessels having an open communication with it.

one side of the wheel, and by the re-action of the valves, successively to give a circular motion to the wheel, the valves opening in the direction in which the weights are pressed, but not in the contrary one, as the steam vessel which moves round it is supplied with steam from the boiler, and that which has performed its office may either be discharged by means of condensers, or into the open air.

"Sixthly, I intend, in some cases, to apply a degree of cold, not capable of reducing the steam to water, but of contracting it considerably, so that the engines shall be worked by the alternate expansion and contraction of the steam.

"Lastly, Instead of using water to render the piston or other parts of the engines air and steam-tight, I employ oils, wax, resinous bodies, fat of animals, quicksilver, and other metals, in their fluid state."

J. Stewart, 1769.

This engine produced a rotative motion by a chain going round a pulley, and also round two barrels furnished with ratchet-wheels, with a weight suspended to the free end of the chain, which thus served to continue the motion of the apparatus during the return of the piston.

M. Washborough, Bristol, 1778.

This invention, like the preceding, was intended to communicate a rotatory motion, without the intervention of a crank. It had a toothed sector on the end of the working beam, acting in a trundle, which, by means of two pinions, with ratchet-wheels, produced a rotative motion in the same direction by both the ascending and descending stroke of the piston; and by shifting the ratchets, the motion could be reversed at pleasure. In this engine Mr. Washborough employed a fly-wheel, which may not unaptly be considered as a magazine of power in this, as in every other species of machinery.

J. Steed, Lancashire, 1781.

This specification was to secure to the patentee the application of a crank for producing continuous motion nearly similar to that now in use.

J. Hornblower, Penryn, Cornwall, July 31, 1781.

The expansive principle of Mr. Watt is in this engine employed by means of two cylinders, by the use of which the force of the apparatus is more nearly equalised. As, however, this invention has been the basis of several improvements of the first magnitude in the introduction of double cylinder engines, we subjoin the specification in Mr. Hornblower's own words.

" First, I use two vessels in which the steam is to act, and which in other steam engines are called cylinders. Secondly, I employ the steam after it has acted in the first vessel to operate a second time in the other, by permitting it to expand itself, which I do by connecting the vessels together, and forming proper channels and apertures, whereby the steam shall occasionally go in and out of the said vessels. Thirdly, I condense the steam, by causing it to pass in contact with metallic substances, while water is applied to the opposite side. Fourthly, to discharge the engine of the water employed to condense the steam, I suspend a column of water in a tube or vessel constructed for that purpose, on the principles of the barometer, the upper end having open communication with the steam-vessels, and the lower end being immersed in a vessel of water. Fifthly, to discharge the air which enters the steam-vessels with the condensing water or otherwise, I introduce it into a separate vessel, whence it is protruded by the admission of steam. Sixthly, that the condensed vapour shall not remain in the steam vessel in which the steam is condensed, I collect it into another vessel, which has open communication with the steam-vessels, and the water in the mine, reservoir, or river. Lastly, in cases where the atmosphere is to be employed to act on the piston, I use a piston so constructed as to admit steam round its periphery, and in contact with the sides of the steam-vessel, thereby to prevent the external air from passing in between the piston and the sides of the steam-vessel.

J. Watt, Birmingham, March 12, 1782.

This invention, which is for an improvement on Mr. Watt's prior patent, consists principally in an advantageous mode of stopping the admission of steam at a given point, so that a part of the working stroke is effected by the expansion of that portion of the elastic vapour which has already entered the cylinder. Several very ingenious contrivances are also described by Mr. Watt for equalising the motion of the piston. The first of these is by a chain acting upon a spiral or fusee; secondly, by levers acting unequally upon each other; and, thirdly, by a large weight attached to the working-beam at a considerable height above the centre of motion. In the last of these methods, when the piston begins its descent, the weight will oppose itself to the motion of the piston, until the descent of the latter have inclined the beam so much, that the centre of gravity of the weight is perpendicularly over the centre of motion of the beam: the weight will then have no effect on the engine; but after it has passed this position, it must evidently tend to aid the effort of the piston to draw up the load of water in the pumps, and render its motion equable.

J. Watt, Birmingham, June 14, 1785.

The object of the present invention is to facilitate the combustion of smoke by a more equable supply of oxygen, and consists in causing the smoke, which is usually emitted on a supply of fresh fuel, to pass, together with a current of air, through the ignited mass that has already ceased to smoke, by which means it will be effectually consumed, and converted into heat or flame. This invention is put in practice, first, by stopping up every avenue or passage to the chimney, except such as are left in the interstices of that part of the fuel which is ignited; secondly, by placing the fresh coal above, or nearer to the external air, than that which is burning, and already converted into coke or charcoal; and, thirdly, by constructing the fire-place in such manner, that the fresh atmospheric air which animates the fire, and the smoke which proceeds from the fresh fuel, must take a downward direction, so as to pass through the whole mass of burning fuel to the most remote part of the fire-place; and by this means the whole of its hydrogen, azote, and carbonaceous matter is usefully employed.

T. Burgess, June 9, 1789.

In Mr. Burgess's apparatus, which was intended to produce a rotatory action, a heavy fly-wheel was set in motion by the alternate elevation and depression of the working-beam. This was effected by an elastic cord passing round a collar on its axis, one end of which was connected with the working-beam, while the other supported a weight. The moveable collar being furnished with a click acting in a ratchet-wheel firmly screwed to the fly-arbor. From this it will be seen that the elevation of the piston would give a proportionate impulse to the fly, which could not be impeded by its subsequent depression, the moveable ratchet-wheel allowing the fly to continue its rotative motion.

Messrs. Bramah and Dickinson, Jan. 15, 1790.

For an engine on a rotative principle.

J. Sadler, Oxford, June, 10, 1791.

The above patent, which Mr. Sadler states in his specification to have for its object, the reducing of the consumption of coals, and consequently the expense of generating steam, appears but little calculated to answer this or any other useful purpose. Mr. Sadler produced a rotatory motion by a hollow cylinder connected with a boiler, which was driven round by the emission of steam from two moveable arms turning upon the same axis.

Francis Thompson, 1793.

A double-acting engine, for turning machinery by a crank. This was effected by employing two cylinders, one inverted over the other; both pistons being connected by one rod, which passed through the upper end of the inverted cylinder, where it was connected with the beam, and thus made a double stroke.

R. Street, Christchurch, Surrey, May 2, 1794.

J. Strong, Bingham, Nots. May 31, 1796.

This patent was obtained for improvements in the piston-cylinders and valves, none of which were ever generally adopted.

W. Batley, Manchester, June 28, 1796.

E. Cartwright, Middlesex, Nov. 11, 1797.

In this engine the condensation is performed by the application of cold to the external surface of the vessel containing the steam. This is effected by admitting the elastic vapour between two metal cylinders, lying one within the other, and having cold water flowing through the inner one, and surrounding the outer one. By these means a very thin body of steam is exposed to the greatest possible surface of cold metal. By means of a valve in the piston, there is a constant communication at all times between the condenser and the cylinder, either above or below the piston, so that whether it ascend or descend, the condensation is always taking place. This mode of condensation also affords an opportunity of substituting alcohol in the place of water, for working the engine, which could not be effected where the injection water mixes with the elastic vapour; and by the employment of ardent spirit, Mr. Cartwright calculates that a saving of half the fuel usually employed might be expected.

T. Rowntree, Blackfriars, May 1, 1798.

J. Hornblower, Penryn, Cornwall, June 8, 1798.

The rotative engine described in this specification, though very ingenious in its construction, is much too complicated to be generally adopted.

J. Dickson, Dockhead, July 14, 1798.

This engine, like the preceding, has never been employed on a large scale, which may be principally attributed to the air-pump and quicksilver, which form an essential part of its construction.

F. Rapozo, Lisbon, Aug. 29.

G. F. Queiroz, Waltham Green, Middlesex, Sept. 1798.

Mr. Queiroz's improvements consist in diminishing the friction, in communicating a circular motion, and in a considerable alteration in the form of the boiler, by dividing it into several compartments, by which a greater surface is exposed to the fire than by the ordinary method, and consequently more steam produced by a given quantity of fuel.

J. Wilkinson, Rotherhithe, July, 1799.

The advantage derivable from Mr. Wilkinson's invention, consists in a closer application of heat to the bottom of the steam boiler: this he effects by constructing it of considerable length, and without any flues round the exterior surface. The heat from the flues, passing uniformly along the bottom, rises up at one end, and returning by flues passing through the water to the other, opens into the chimney which carries off the smoke.

M. Murray, Leeds, July 16, 1799.

With a view to save fuel, Mr. Murray provides the top of the boiler with a small cylinder and piston, connected with a rack, by means of which the steam within the boiler raises up or depresses a damper in the chimney. He is thus enabled by the increase or decrease of the steam to keep up a due proportion between the elasticity of the vapour thus generated and the draught of the fire. Mr. Murray's second improvement consists in placing the ordinary steam cylinder and piston in a horizontal, instead of a vertical direction, and by this means a much longer stroke may be obtained than in the usual way. He also causes the pistons, by their reciprocating motion, to produce a circular motion of equal power, and enables the engine to fix the wheels necessary for producing motion alternately in perpendicular or horizontal directions.

W. Murdock, Redruth, Cornwall, Aug. 29, 1799.

Mr. Murdock describes his improvements under four heads, viz. First, a more equable mode of boring the metallic cylinders and pumps by means of an endless screw worked by a toothed wheel. Secondly, by simplifying the construction of the steam-vessel and steam-case in engines formed on Mr. Watt's principle, which he effects, by casting the steam-case of one entire piece, to which the cover and bottom of the working cylinder are attached. He likewise casts the cylinder and steam-case in one piece of considerable thickness, and bores a cylindric interstice between the steam-case and steam-vessel, leaving the two cylinders attached at one end, and he closes the other by a ring of metal.

Thirdly, he simplifies the construction of the steam-valves or regulators, in Mr. Watt's double engine, by connecting together the upper and lower valves so as to work with one rod or spindle. The steam or tube which connects them being hollow, serves as an eduction pipe to the upper end of the cylinder, and a saving of two valves is effected.

Fourthly, by the assistance of two cogged wheels working in an air-tight vessel, a rotative engine of considerable power is produced, though from the great difficulty attached to the making it air-tight, and the waste of steam in condensation, the plan does not appear of much practical utility.

J. Bishop, Covent Garden, Sept. 23, 1799.

This apparatus, which may be considered as a modification of Amonton's fire-wheel, consists of three parts: First, a wheel similar to an overshot water-wheel; secondly, a large close vessel or case, made of copper or iron, in which the wheel is fixed, and which is to work in a vertical direction; and thirdly, an air or steam-pipe and valve, which passes out at the top of the close vessel, through a small reservoir of cold water, for the purpose of condensation, if necessary.

Motion is given to the wheel, and to any connecting machinery, by the power of ascending steam, in the following manner: the close copper vessel in which the vertical wheel is fixed, is filled with water above the axle of the wheel. The application of fire at the bottom of the vessel will raise the steam; and the steam in its ascent, by entering into and acting upon the bucket of the wheel, will give it a rotatory motion, producing a power of about nine pounds for every gallon of water employed. A contrivance is also added for the production of a vacuum by the means of an air or steam pipe, by which a considerable reduction in the expenditure of fuel is effected.

P. Crowther, Newcastle-upon-Tyne, Feb. 28, 1800.

J. and J. Robertson, Birmingham, Aug. 13, 1800.

Messrs. Robertson's invention consists in preventing, in a great measure, the escape of steam which usually takes place even in engines of the best construction by the wear of the materials of which the piston is composed, and by an apparatus exclusively the invention of the patentees, the small portion of elastic fluid that finds a passage during the action of the cylinder is employed on another piston, thus augmenting the power of the engine. These desiderata are stated to be effected by the use of two steam cylinders, one smaller than the other, with pistons fitted to each, and whose united action is described in the above specification.

The patentees have also effected a considerable saving of the fuel usually employed, by an improvement which evidently possesses very considerable advantages.

The coal is admitted into the furnace by a hopper or mouthpiece, so that it may fall into the fire-place above the bars, as the fuel is expended. From the upper side of the hopper a stream of fresh air rushes downwards on the fire, and by the adoption of this plan a large portion of the smoke is converted into flame.

E. Cartwright, St. Mary-le-Bone, Feb. 5, 1801.

The engine here described was never, we believe, fairly tried, at least judging from the simplicity of its parts, capability of control, and great portability, it is not too much to suppose, had its merits been better known, it would (with some modifications) have come into more general use.

W. Hase, Saxthorpe, Norfolk, May 14, 1801.

After leaving the cylinder in order to be condensed, the steam is in this apparatus conducted through a vessel containing a number of metal pipes filled with water from the condenser. This being surrounded by the steam, imbibes a portion of its caloric, while it facilitates the process of condensation. The water thus heated is immediately conveyed to the boiler, which is preserved at the boiling point by a small addition of caloric.

M. Murray, Leeds, August 4, 1801.

This patent comprises six principal objects: First, by an improved air-pump, the gaseous matter is discharged from the cylinder without any effort in opening of valves, or pressing through a body of water; and it also causes the water and air to be discharged separately and by different ways: this is effected by taking out the air alone by a bucket, and the water alone by another, or by an eduction-pipe twenty-eight feet long. The second principle is an improved method of packing the cylinder's lid, stuffing-boxes, &c. by bringing the moveable parts of each in immediate contact; this is effected by placing the necessary packing on the upper side of the cylinder lid, which prevents the piston-rod receiving friction from any oblique pressure, by the lid being screwed down more upon one side than the other.

The two next improvements relate to the construction and circular motion of the valves, the uppermost two being inverted; and the valve-rods are made to pass through the reservoirs of oil, or other fluid matter, which effectually prevents the air from insinuating itself into the engine.

The next principle is a new method of connecting the

piston-rod to the parallel-motion; and the last relates to the construction of fire-places, by which the smoke arising from the fire is consumed, and made part of the fuel; on this latter head, however, Messrs. Brunton, Parkes, and Losh, have made considerable improvements, so much so, indeed, as to supersede the application of this clause in Mr. Murray's patent.*

J. Bramah, Pimlico, Nov. 28, 1801.

In this engine Mr. Bramah employed a four-passaged cock for the emission of steam from the boiler, which in this case is made to enter into a hollow at the large end of the cone of the cock, and to pass away to the condenser by a passage at the smaller end of the cone of the cock. By this means the metallic fitting is always rendered perfect, the plug being pressed into its seat by the force of the steam, acting upon a surface equal to the small end of the cock, from which the pressure is relieved. Mr. Bramah, also, makes his four-way cock turn continually in the same direction, by which means the same effects are produced as by turning it backwards and forwards, but the wear is rendered more equable.

W. Symington, Kinnaird, Stirlingshire, Oct. 14, 1801.

For a rotatory and other motions, without the interposition of a lever or beam.

J. Sharper, Bath, Jan. 28, 1802.

R. Trevithick and A. Vivian, Camborne, Cornwall, March 24, 1802.

The high-pressure engines of Messrs. Trevithick and Vi-

* This patent was, we believe, set aside by a writ of *scire facias* instituted by his Majesty's Attorney General, at the instance of Messrs. Boulton and Watt, who had previously practised some things contained in Mr. Murray's specification.

vian, were expressly intended for the propelling of carriages upon rail roads. When employed for this purpose the boiler was composed of cast iron of a cylindrical form. This was mounted horizontally upon four wheels, the cylinder of the engine being placed vertically in the end. Two connecting rods, descending from the cross bar of the piston, were then made to communicate motion to the wheels by means of a crank; no fly-wheel being necessary, the momentum of the carriage carrying the cranks past the lines of the centre.

M. MURRAY, Leeds, Yorkshire, June 28, 1802.

This patent, which was for a portable engine, combining some of the most useful of Messrs. Boulton and Watt's inventions, was at their instance repealed in the following year.

T. SAINT, Bristol, Dec. 21, 1802.

The principle of this invention may be thus described: at the bottom of the boiler an opening is made nearly as large as the flue; on this opening is fixed a tube, through which a communication takes place between that part of the fire-place in which the flame rises or circulates, and the interior space of the boiler in which steam is produced for the supply of the engine. This aperture may remain open, but Mr. Saint recommends the application of a safety-valve so placed that no part of the heated air contained within the boiler shall be permitted to escape through the opening, but that the flame from the furnace may be admitted as often as a reduction in the elasticity of the compound steam will admit.

M. BILLINGSLEY, Dec. 22, 1802.

The usual mode of perforating and finishing cylinders, through the agency of an horizontal apparatus impelling the borer in a vertical direction, is upon many accounts inconvenient, the sand and borings occupying one side of the cylinder, and wearing away the edges of the cutter. By the application of Mr. Billingsley's apparatus, this process is reversed by causing the borer to revolve in an horizontal direction, and thus allowing the sand &c. to fall freely to the lower opening in the cylinder. In this method, the finishing part of the cutter is employed upon a clean face of metal, and not being encumbered with the cuttings, the borer goes completely through, without any attention being necessary.

J. Leach, Merton Abbey, April 7, 1803.

A. Woolf, Wood Street, Spafields, July 29, 1803.

An account of the above patent for an improved mode of converting water or other fluids into steam, will be found prefixed to the description of Mr. Woolf's engine.

B. Donkin, Dartford, August 3, 1803.

A rotatory motion is here produced by the application of steam below the surface of a vessel of water, containing a bucket-wheel, the elastic fluid acting upon each bucket in succession. These, when filled with and rendered buoyant by the steam, will ascend with considerable force, carrying with them any other moveable apparatus to which they may be attached.

W. Freemantle, Hoxton, Nov. 17, 1803.

The first improvement described by Mr. Freemantle in this specification, consists in making the steam cylinder or cylinders in such a manner that the diameter of the bore shall be about equal to the length of the stroke, by which means the friction will be considerably reduced.

Another part of Mr. Freemantle's invention is an universal

circular valve, which, when the engine is at work, vibrates on its axis forty-five degrees each way, and by its alternate oscillations admits the steam uniformly above and below the piston. When the steam is admitted into the top of the cylinder, the bottom communicates with the condenser; and *vice versa.*

When the steam is only applied to the bottom of the piston, as in the atmospheric engine, two steam cylinders are to be employed instead of one, and by this means the force of the engine is so nearly equalized as to act with a much smaller fly than is usually employed. In the latter case, however, the pistons of the cylinders are made to act on the two arms of a crank, placed at right angles to each other; and the valve is not to vibrate but revolve on its axis, making one turn to each stroke of the engine.

The parallel motion employed in Mr. Freemantle's engine is next described: this he effects by the application of a compound lever.

Another improvement is in the cold-water pump; this consists in placing an air-vessel so as to communicate with the ascending column of water immediately below the lower valve of the pump. Now it is evident, that before water can be raised into the barrel, a partial vacuum will be formed in the air-vessel, the amount of which will depend on the height of the vessel above the surface of the water in the well. In pumps of the ordinary construction it sometimes happens, that the velocity of the piston exceeds that of the water: hence a violent concussion is produced highly detrimental to the pump and connecting apparatus; this is prevented by the adoption of Mr. Freemantle's improvement; for, instead of a vacuum being formed below the piston, air will be extracted from the air-vessel; and, as the piston descends, the pressure of the air on the surface of the water in the well not being counterbalanced, will continue to rise in the vessel till the equilibrium is restored; and when the piston ascends again, the barrel will not only be supplied by the pipe, but also by the air-vessel.

R. Willcox, Bristol, April 30, 1804.

The improvement for which this patent is taken out, consists in lessening the consumption of fuel by the application of vapour in a high state of elasticity; and in an addition to the chimney flue of a furnace, by which the descent of the smoke and heated matter to a lower level than that of the fire-place is regulated and adjusted at pleasure.

As the caloric which escapes by the chimney in various manufactories is very considerable, and, according to Mr. Willcox, more than sufficient to work an ordinary steam engine by condensation, he finds it most advisable to load the safety valve, and construct the engine of the requisite strength to bear an elastic action of from fifteen to one hundred and fifty pounds on the inch; and in this manner the machine is worked by its elasticity only. By this application of the steam, the patentee states that the power of a four-inch cylinder may be made to equal that of one twelve times its diameter.

Among the advantages derivable from the use of this engine, it is said that the constant heat of the cylinder and the pipe that leads to it, which can never come in contact with the cold water, and the increasing heat of the water in the cylinder, which soon acquires a high temperature, and then continues its place, by its diminished specific gravity, must tend in the highest degree to prevent a wasteful condensation of steam.

A. Woolf, Wood Street, Spafields, June 7, 1804.

This engine is in many respects similar to Mr. Hornblower's, with the addition of employing steam of a high pressure, and in proportioning the capacities of the two cylinders to the increased expansibility of the elastic fluid, according to his table. Mr. Woolf, in his specification, states that he has ascertained by actual experiment that steam acting with the expansive force of four pounds upon the square inch, against the safety valve, exposed also to the weight of the atmosphere, is capable of expanding itself to four times the volume it then occupies, and would still be equal to the pressure of the atmosphere; so, in like manner, quantities of steam of six, seven, eight, &c. pounds the square inch, can expand themselves to six, seven, or eight times their volume, and still be equal to the atmosphere, or capable of producing a sufficient action against the piston of a steam engine, to produce the upward action in Newcomen's atmospheric engine.

An engine constructed on Mr. Woolf's plan, must have two steam vessels of different dimensions, according to the temperature, or expansive force, to be communicated to the steam. Each steam vessel should be furnished with a piston, fitting air-tight, and the smaller cylinders should have a communication both at top and bottom with the boiler which supplies the steam, capable of being opened and shut during the working of the engine. The top of the small cylinder having a communication with the bottom of the larger cylinder, and the bottom of the smaller with the top of the larger, with proper means to open and shut these alternately by means of cocks and valves. A communication should also be formed by the same means between the larger cylinder and a condensing vessel, into which a jet of water is admitted to render the condensation more complete.

When the engine is at work, steam of high temperature is admitted from the boiler, to act by its elastic force on one side of the smaller piston, while the steam which had last moved it has a communication with the larger steam vessel, now moving towards that end of its cylinder which is open to the condensing vessel.

If both pistons end their stroke at one time, and both are placed at the top of their respective cylinders, ready to descend, then the steam entering above the smaller piston will carry it downwards; while the steam below it, instead of being allowed to escape into the atmosphere, will pass into the larger cylinder above its piston, which will take its downward stroke at the same time with the piston of the smaller cylinder. Both pistons having thus reached the bottom of their respective cylinders, the steam is to be shut off from the top and admitted to the bottom of the smaller cylinder, and the communication opened between the top of the smaller and the bottom of the larger cylinder; the steam which, in the downward stroke of the engine, filled the larger cylinder, being now open to the condenser, and the communication between the bottom of the larger cylinder and the condenser shut off, and so alternately admitting the steam to the different sides of the smaller piston, while the steam last admitted into the smaller cylinder passes alternately to the different sides of the larger piston, the top and bottom of which are made to communicate alternately with the condenser.

J. Rider, Belfast, March 26, 1805.

The improvements described in this specification consist, first, in lining the steam-cylinder with a soft metal, similar to pewter, of a sufficient thickness to admit of finishing its inner surface by draw-boring, &c.; secondly, in applying a hollow piston-rod, answering the purpose of an eduction-pipe; and thirdly, in the order of opening and shutting the valves. The last and most important part, however, of Mr. Rider's invention requires a more particular description.

Upon an horizontal arbor, which may be denominated the main arbor, are placed three wheels, a drum or barrel, and a pinion : one of these wheels, that is to say the main wheel, is fitted by means of a socket upon the main arbor, so as to revolve upon its axis, and has teeth both upon the exterior and interior periphery of its rim. Within the circle of the interior cogs of this wheel a pinion is fixed to the arbor, its diameter being one-third of the interior diameter of the main wheel. The moveable barrel turns freely upon the main arbor, and it carries a cord, with a weight hanging at its end similar to a clock-weight. Besides this the ends of the barrel are pierced with two orifices, each at about half the exterior radius of the main wheel from the arbor; these apertures serving as pivot-holes, wherein an arbor turns, carrying a wheel, of which the diameter and number of teeth are equal to those of the pinion : the latter wheel may be called the barrel-pinion ; its teeth work in the teeth of the pinion, and also in the interior teeth of the main wheel. By these means the barrel may be turned round upon the main arbor, while the arbor itself is turned by the pinion. The exterior teeth of the main wheel turn the pinion of a scapement-wheel and pallets. Near one end of the main arbor there is a ratchet-wheel and click ; and near the other end a wheel, which is acted upon by an endless screw upon an horizontal shaft, worked by the usual motion of the engine. This arrangement serves to regulate the rate of the engine's motion ; for the turning of the worm-wheel, last described, causes the weight to be raised which hangs to the cord winding upon the barrel ; and this weight is connected to one end of a lever, the other end of which is attached to the steam valve, so that its elevation depends upon the height to which the weight is raised. The aperture of this valve is formed like an inverted cone ; and while this valve shuts and opens twice at every stroke, the lever does not prevent such opening and shutting, but merely limits the extent of the opening by the action of a rod connected with it; so that when the weight is highest, the valve is least opened, and *vice versa.* Little power is lost by these means, and the speed of the engine can be accurately regulated by adjusting the length of the pendulum to the arrangement of teeth in the wheels and pinions.

J. Hornblower, Penryn, March 26, 1805.

This steam wheel which differs considerably from Mr. Hornblower's prior patent for a rotatory motion, consists of three principal parts: The external circular case, which is shaped like a globe, from which about forty degrees are cut off at each pole; secondly, a partition which divides the case into two parts transversely in the plane of its axis; and thirdly, the moveable or circulating parts which are analogous to the piston in the common steam engine. To form an idea of these last, we must conceive an hollow nave attached to an horizontal axle, which nave is pierced with two pair of circular holes on its cylindrical surface, each pair corresponding at opposite sides to each other; through these holes pass radii, moveable round their own axis for one-half a circuit, to which are attached flat quadrants, placed so that the planes of those at opposite sides of the nave should be at right angles to each other. By this arrangement, when one of them is placed so that its plane shall be at right angles to the axis of the nave, the plane of the other will coincide with that of the axis. In the partition which forms the second principal part of the engine, are two cavities at opposite sides of the centre, one of which corresponds exactly to the shape of the greatest surface of the quadrant; the other is much smaller, and only admits the quadrant edgeways; these cavities are continued in a sort of case of the same shape, for one-fourth of the circle on each side, by which means there are always two of the quadrants, or the greatest part of them, working in each cavity at the same time. The quadrants are made hollow so as to admit of their being stuffed at their edges, as is also the recess in which they circulate.

From this it will be seen that the external circular case is divided by the partition and the quadrantular pistons into two separate chambers, each steam tight; into one of these chambers a pipe is conveyed from the boiler, and from the other chamber another pipe communicates with the condenser; now as the quadrant which occupies the greater cavity of the partition opposes a much greater surface to the pressure of the steam than that which lies edgeways in the smaller cavity, it will be forced forward towards the cold chamber, in which, when it is arrived, it meets a sloping block so shaped, that in passing it it is turned round one quarter, or so as to be at right angles to its former position, and thus enters the smaller cavities edgeways, while by the same movement the opposite quadrant is turned flat across the entrance of the larger cavity of the partition, and on entering it is impelled in its turn by the steam, as before described. The axis, or arbor, on which the nave is fixed, which sustains the quadrants, passes through the outer case in an horizontal direction, and to its extremities are to be fixed those wheels which are to give motion to the machinery required to be worked by the steam-wheel. The outer case is fixed in a vertical position, and has a flat plate cast at the part intended to be lowest, by which it may be bolted to the floor of the building in which it is erected. It is formed so as to separate into three horizontal sections, the middle one of which is for the purpose of admitting and properly fastening the partition with its stuffing boxes; the upper section serves as a lid, and all are secured to each other by flanches and screws.

W. Earle, Liverpool, March 26, 1805.

J. C. Stevens, May 31, 1805.

This patent is for a high-pressure boiler, resembling Mr. Woolf's; it may be thus described. Suppose a plate of brass, of one foot square, in which a number of holes are perforated, into each of which is fixed one end of a copper tube, an inch in diameter, and two feet long, and the other end of the tube inserted in like manner into a similar piece of brass: these are to be enclosed at each end of the pipes by a strong cap of metal so as to leave a small space between the plate and cap. The necessary supply of water is then to be injected by means of a forcing pump communicating with one of the caps, while the steam is conveyed from the other to the required point.

A. Brodie, May 31, 1805.

The cast-iron boiler invented by Mr. Brodie, may be constructed of any dimensions, and the iron plates of which it is composed are made with flanges of the required size, and put together with rivets and screws. To prevent the boiler giving way by the force of elastic vapour, it is strengthened by wrought-iron stays, and the vessel thus made is supported by iron legs, so that the fire is allowed to communicate with the whole of its lower area without being connected with the brick work with which it is usually surrounded.

James Boaz, Glasgow, July 2, 1805.

This patent is for an improvement on Savery's engine, which Mr. Boaz effects by separating the steam from the water to be raised. For this purpose he employs a floating piston upon Papin's plan, and such an arrangement of the forcing pipes that the repellent force of the steam is always acting upon the same body of water.

A. Woolf, Spafields, July 2, 1805.

To prevent the danger attendant on the use of Mr. Woolf's high-pressure engine, he in this patent recommends the use of oil, or the fat of animals placed in the receptacle, which in his former patent contained steam of great elasticity. The vessel employed to contain these fluids forms a complete case or envelope to the working cylinder, so that the whole is maintained at one uniform temperature, which is to be kept up by a fire under or round the receptacle. By this arrangement, the necessity of employing steam of a great expansive force is obviated, and steam of a comparatively low temperature will produce all the effects that can be obtained from steam of a high temperature, without any of the risk with which the production of the latter is accompanied. He also proposes a method of preventing the passage of any of the steam from that side of the piston which is acted upon by the steam, to the other side which is open to the condenser. In the double-action steam engines he effects this by employing, upon or about the piston, a column of mercury of an altitude equal to the pressure of the steam. In working the single engine, a less considerable altitude of metal is required, because the steam always acts on the upper side of the piston; and in this case oil, or the fat of animals, will answer the purpose sufficiently well, and at much less cost.

W. Deverell, Blackwall, August 2, 1805.

This specification describes an improved construction of the fire-place, an improvement in the cold water pump, and a saving in the method of applying the steam. The principal of these is that of connecting the steam-boiler with three iron cylinders, filled with water instead of placing it on brick work. In describing the peculiar advantages resulting from this part of the invention, Mr. Deverell says, "In the present mode of setting boilers, the brick work underneath them is attended with frequent repairs, owing to the action of the fire upon them; nor is this expense the only inconvenience: the whole concern is, for the time that they are repairing, completely stopped." These inconveniences Mr. Deverell proposes to remedy by the adoption of metallic supports for the boiler.

As a more economical mode of applying the steam, Mr. Deverell, like Mr. Woolf, proposes to have two working cylinders placed near to one another, each having a pipe of communication, with a large vessel in which the steam, after passing from the small cylinder, is suffered to expand itself before entering the large cylinder. The pistons in the two cylinders work alternately up and down by means of valves or cocks, opening and shutting as in the common engine. Suppose the small piston has made a stroke, and a passage is opened to the steam vessel at the end of the stroke; at first beginning to work the engine the vessel will be full of steam of about eighteen pounds pressure, admitted from the boiler, but afterwards will only be supplied by the steam thrown in from the small cylinder. If the steam in the boiler be of forty-four pounds pressure per square inch, the ratio of the two working cylinders may be as one to three, for then the smaller one will supply the larger with steam of about eighteen pounds pressure.

The improvements thus effected consist in a saving of the pressure of the atmosphere, and in the steam which would otherwise be discharged and useless going from the smaller working cylinder to the steam vessel, and from thence to the larger working cylinder, from which it is afterwards drawn off and condensed.

S. Miller, Gresse Street, St. Pancras, Oct. 30, 1805.

J. Trotter, Soho Square, Nov. 14, 1805.

A. Flint, Northampton Street, Nov. 16, 1805.

This engine, which may be used either as an hydraulic machine, and impelled by a continuous stream of water, or as a rotative steam-engine, consists of two hollow cylinders, one of which is so much smaller than the other that it may lie within it. They are both to be turned true, and placed concentrically: they are also furnished with flat steam-tight tops and bottoms, either cast with them, or fastened by screws. The inner cylinder has a partition in its middle parallel to its top. It revolves within the outer cylinder, and has a pipe passing through the centre of its top and that of the outer cylinder, from its upper division, in the latter of which tops it is made steam-tight by stuffing boxes. This pipe communicates with another, that passes to the boiler, having the parts in contact with it also made steam-tight so as to admit of its circular motion. Another pipe, in a similar manner, passes from the lower division of the inner cylinder through the bottom of the outer cylinder, to form a connection with the condenser.

From the side of the inner cylinder projects a piece similar to a piston, which fills the section of the cavity in the line of the radius, between the two cylinders: this rectangular piston is so contrived, that it may be stuffed round the edges and be made steam-tight.

The outer cylinder has two semi-cylindrical cavities cast in its sides opposite to each other, with their open parts turned inwards; each of which is of sufficient size to admit within it a portion of a smaller cylinder, which is placed upright between the two large cylinders. In these spaces are placed valves, which retire alternately into the lateral cavities of the outer cylinder to admit the piston to pass by them. These valves consist each of the segment of a cylinder of the height of the inner cylinder, connected with a circular top and bottom, turning on centres; and an axle from each passes through the top of the outer cylinder, through steam-tight joints, by which it may be turned round from without.

At one side of the piston a perforation is made into the upper cavity of the inner cylinder to admit the steam; and at the other side of the piston a similar perforation is made into the lower cavity of the inner cylinder to form a communication with the condenser. An arm also projects from the revolving steam-pipe, which as it moves round strikes against other arms projecting from the axles of the valves, and opens them in succession, while connecting rods passing between the arms of the valves and other arms, are so arranged as to close one valve when the other is opened. The steam being now admitted will pass on from the steam-pipe through the upper cavity in the inner cylinder, to the space intercepted between the two large cylinders, the shut valve, and the piston, and will impel the piston round till it has passed the open valve; after which the revolving arm before mentioned closes the open valve, and opens the shut valve, which operation is successively repeated. In the mean time the confined steam enclosed in the space first mentioned, escapes at the opening of the valve into the lower cavity of the inner cylinder, and from thence to the condenser, and thus maintains that inequality of pressure at the opposite sides of the piston which causes it to revolve.

R. Willcox, Lambeth, May 21, 1806.

Mr. Willcox's improved steam engine consists of an outer fixed cylinder, and an inner revolving one, each furnished with pallets or cocks, which in passing each other are moved so as to recede from each other's way; but in other parts of the circle they project so as to traverse the space between the two cylinders, and form steam-tight partitions, one of which being fixed, and the other moveable, the steam forces the latter round, and with it the moveable cylinder, the axis of which gives motion to the machinery for which the engine is erected. On one side of the fixed pallet is a valve, which by a pipe communicates with the boiler, and on the other side is placed a second valve which leads to the condenser. The ends of the cylinders are made steam-tight by rings which press the packing against them; and the edges of the pallets are made steam-tight by the intervention of a hempen cloth. That part of the surface of the cock which comes in contact with the revolving cylinder, has a groove cut down it, into which a piece of metal is fitted, that is pressed against the cylinder by screws, so as to come in close contact with the revolving cylinder. In cases where this engine is employed to raise or give motion to any fluid introduced within its interior chamber, the effects produced will be similar to those of a lifting or forcing pump, and it is likewise applicable to all engines which operate by giving motion to fluids.

R. Dodd, Change Alley, London, June 6, 1806.

W. Nicholson, Soho Square, Nov. 22, 1806.

The method in which steam is directed to be applied in the specification of this patent, is similar to that in which water acts in the ancient instrument called the water-blast, and in the same manner will impel forward air, or any other gaseous substance, in contact with the perforations of the tube through which it passes. The apparatus employed by

Mr. Nicholson consists of a boiler, from the top of which a horizontal tube passes in a direction perpendicularly over the centre of an air receptacle, when it bends downwards for a small distance, that the current of steam proceeding from it, may enter a vertical pipe beneath it, the lower end of which passes a little way underneath the water, with which the air receptacle is about half filled. The use of the water in the air receptacle is by condensing the steam to separate it from the air, so that the latter may pass on freer from any aqueous mixture than it otherwise would. A way is described of applying this operation of steam, in forcing air forward, to aid the water-blast, in which the water is made to pass through the side apertures of the descending pipe from an external vessel, while the air pressed forward by the steam, passes down a pipe which enters a small way into the upper part of the same descending pipe.

H. Maudslay, Margaret Street, Cavendish Square, June 13, 1807.

These improvements, as will be seen by reference to the plate and description, consist in reducing the number of parts in the common steam engine, and so arranging and connecting them as to render it more compact and portable. This is effected by employing metallic frames, beams, &c. instead of wood, brick-work, and other materials usually applied to that purpose.

T. Preston, Tooley Street, Borough, Jan. 26, 1808.

A new method of setting boilers.

T. Smith, Bilston, Staffordshire, June 3, 1808.

T. Price, Bilston, Aug. 24, 1808.

T. Mead, Yorkshire, Aug. 24, 1808.

This patent is for a rotative engine, and the inventor employs two moveable pistons which alternately revolve round their axes or centres. To effect this, two circular plates or shells of metal are made similar in their construction, each of which has a flanch and a semi-circular cavity formed for the reception of the pistons. A hollow part is also formed round the centre of each for a small circular plate to turn in; and near the edge of this recess a small groove is made containing the requisite packing wadding, &c. On the outside of each of the metallic shells there is a hollow pipe or boss for the reception of two spindles that pass through them. Two holes are also made through one of the plates for the insertion of pipes, one for the purpose of conveying steam into the shells, and the other for conducting it from them into a condenser. One of the shafts or spindles is made hollow, to permit the other to pass through it. To the lower part of each of these shafts a piston is fixed; and each has also attached to its upper part an arm with a friction-wheel near its outer extremity. When the pistons are put in motion the friction-wheels work in, and communicate motion to the fly, and other apparatus to be impelled.

J. P. Fesenmeyer, St. Clement Danes, June 15, 1809.

E. Lane, Shelton, Staffordshire, Aug. 9, 1809.

W. C. English, Twickenham, Nov. 28, 1809.

W. Noble, Battersea, Dec. 14, 1809.

S. Clegg, Manchester, July, 26, 1809.

For a rotative engine, the piston of which makes a complete

revolution in a channel at a distance from the centre of motion. Although this apparatus is exempt from some portion of the friction inherent in engines on the rotative construction, it still remains more considerable than Mr. Clegg seems to suppose. The difficulty also of making such an extent of surface air-tight, and the liability of some of its parts to derangement, appear great drawbacks upon its utility.

R. WITTY, Kingston-upon-Hull, Feb. 12, 1810.

This invention consists in making, arranging, and combining the reciprocating rectilinear motion with the rotative in such a manner that the steam-cylinders, with pistons moving in them in a rectilineal direction, do at the same time turn upon a horizontal axle or shaft. To effect this the patentee employs four cylinders fixed at right angles to each other on a hollow nave or axle, by means of screw-bolts; and the pistons working in these cylinders are made tight at their extreme ends by the usual packing. These pistons, which are firmly connected together in pairs, by reciprocating rods, must be made of such a weight, that a vacuum in one of the cylinders may easily raise them both together in a perpendicular direction. An axle which is fixed horizontally is ground air-tight into the hollow nave, like the key of a cock, with two ducts or tubes in it; one of these tubes is placed at the upper side of the axle, and is connected with the steam-pipe; the other is fixed on the opposite side, and joined to the pipe that leads to the condenser. Each cylinder is made to communicate through the hollow shaft where the two ducts in the fixed axle (which resemble two water ways in a cock) correspond with each other, and at each half revolution the holes in the bottom of the cylinders open alternately into these two ducts. The hollow shaft must be made of sufficient length on each side of the wheel to admit of being supported in brass pivot holes. The cylinders and pistons being thus arranged, and one pair of them being nearly in a vertical position, if the steam be admitted into the upper cylinder by the proper duct, its expansive force will raise this pair of pistons, and thus destroy the equilibrium of the wheel, producing a rotatory motion. In revolving, each of the cylinders will be filled with steam from the upper duct, and discharged when they descend to the lower one. Thus after the cylinder has cleared itself of air at the commencement of its operation, the lower cylinder will be under a vacuum, while the upper one connected with it, will be receiving steam from the boiler. Hence the pistons will evidently be constantly receding from the centre on one side of the wheel, and approaching the centre on the other, thus producing a continuous motion.

A. Woolf, Lambeth, June 9, 1810.

The working cylinder for this engine has no bottom, but is enclosed in another cylinder of such dimensions, that the space between the two is equal, at least, to the contents of the working cylinder, the lower rim of which is about the same distance from the bottom of the enclosing cylinder as the sides of the two cylinders are apart.

Into the enclosing cylinder such a quantity of oil or fat is put, as shall, when the piston is at its greatest height in the working cylinder, fill all the space beneath it, and also fill the enclosing vessel to the height of a few inches above the lower rim of the working cylinder. A small quantity of oil is also poured in above the piston. If the engine is to be open to the atmosphere, the enclosing vessel has a communication with the boiler above, which, when opened, causes the oil or fat to ascend beneath the piston as it rises; and when the passage to the boiler is closed, and that to the condenser opened, the pressure of the atmosphere forces the oil back again into the enclosing vessel. In a closed or double engine, the communications from the boiler and condenser are to be to the top of the working cylinder, and to the bottom of the enclo sing vessel. In the event of evaporation, means are to be provided by cocks, or valves, and a spring-pump, to keep the oil at a due height over the piston. By thus interposing oil between the piston and steam, both above and beneath, a considerable saving of steam, and consequently of fuel, is effected.

R. Witty, Kingston-upon-Hull, Oct. 30, 1811.

This specification describes several improvements on a prior patent obtained by Mr. Witty. They consist principally in making the piston draw or force round the machinery to be worked by it, whilst itself moves both in a rectilinear and rotatory direction in a cylinder or steam-vessel, which also revolves upon an axis.

To admit the action of the steam, and of the condenser, in the revolving cylinder, its axis is bored lengthways in two places, so as to form two passages, each of which communicates by lateral pipes with the end of the cylinder opposite to the side of the axis in which it lies; the extremity of this perforated axis is formed of a conical shape, and turns in a box made to fit it, in the same manner that the revolving part of a common cock turns in its barrel. From the upper part of this box a pipe passes to the steam-boiler, and from the lower part another pipe proceeds to the condenser, and lateral apertures are made through the sides of the axle to the two passages within it, which, as the axle turns, alternately communicate with the steam-pipe and the pipe of the condenser in the box, in the same manner as a two-way cock is made to act.

Several principles are mentioned by the patentee, on which the cylinder prepared as above, can force itself round; which are all of the nature of crank or cardioid motions. The first of these principles stated by the patentee effects a rotatory movement by the action of a moving groove on a fixed centre; which groove is placed at right angles to the cylinder, in a frame that is connected with piston-rods proceeding from the opposite ends of the cylinder, and of course partakes of their alternating motion. Another principle consists in the operation of the ends of piston-rods, proceeding from the opposite extremities of the cylinder, on the outside of the rim of a large wheel, whose centre is placed at the distance of about half the stroke of the piston from the axis of the cylinder. The rim of the wheel projects so as to extend to the line of the piston-rods, which are bent round to support friction wheels outside it, that alternately come in contact with steps on the rim, and by them force round the wheel, by a motion similar to that which levers would cause, when made to press alternately on the outside of a heart-wheel.

C. BRODERIP, Great Poland Street, Nov. 2, 1811.

R. W. FOX and J. LEAN, Budock, near Falmouth, Dec. 10. 1812.

The two principles on which the patentees profess to have founded their improvements upon furnaces, are, First, That an artificial blast will produce an equal quantity of heat with one-fourth less fuel than the usual open draught; and, Secondly, That the separation of that part of the boiler, which is immediately over the fire, will permit steam of a higher temperature to be collected, which might be advantageously applied in its passage to a lower temperature. In this construction the air for the consumption of the fuel is not permitted to enter the lower parts of the fire-place in the usual way, but is injected or forced through openings by machinery attached to the steam engine.

W. CHAPMAN, Murton House, Durham, Dec. 30, 1812.

In this patent a moveable chain is employed to impel carriages upon a plain road instead of the crank usually applied to the carriage wheel. Mr. Chapman also employs an additional number of wheels for the support of the carriage, by which means a considerable saving in the cost of the rail-way is effected.

W. BRUNTON, Butterley, Derbyshire, May 22, 1813.

The pedestrian apparatus or walking machine described by Mr. Brunton, is constructed to obviate the necessity of employing an iron rail or carriage way. This he attempts to effect by the use of metal legs occasionally raised and depressed by the power of a steam engine similar to the motion of the human frame, when in the act of walking. That this however is not considered sufficient, even by the patentee, we have abundant proof, and the specification contains a description of various modes of connecting the machine with an indented rail, &c. any of which are in the full as expensive as the common cogged wheel and tooth track road employed by Mr. Blenkinsop.

R. DUNKIN, Penzance, Jan. 30, 1813.

For lessening the consumption of steam and fuel.

R. WITTY, Kingston-upon-Hull, June 5, 1813.

J. BARTON, Tufton Street, Westminster, Nov. 1, 1813.

J. WHITE, Leeds, Dec. 14, 1814.

W. A. NOBLE, Riley Street, Chelsea, March 23, 1814.

J. U. RASTRICK, Bridgnorth, Salop, April 1, 1814.

T. Tindall, York, June 18, 1814.

For improvements in the application of steam to the propelling of carriages.

R. Dodd and G. Stephenson, Killingworth, Northumberland, Feb. 28, 1815.

For improvements in the construction of locomotive engines.

W. Losh, Northumberland, April 8, 1815.

The first object to be attained by the adoption of this patent is a considerable saving in the consumption of fuel employed in heating the boiler. This the patentee proposes to effect by employing two furnaces, and thus preventing the usual current of cold undecomposed atmospheric air from passing along with the heated gaseous matter. When the atmospheric air in this undecomposed state comes in contact with iron, and some other metals at a high degree of heat, it has the effect of oxidating their surfaces, both by its own decomposition, and by that of the water which it always carries with it; and those oxidated surfaces separate in successive coats of scales, till by degrees, the metal is entirely corroded away, so that by the adoption of Mr. Losh's plan, a considerable saving in the expense of repairs and loss of time in replacing the boiler is effected.

M. Billingsley, Bradford, Yorkshire, April 20, 1815.

R. Trevithick, Cambrone, Cornwall, June 6, 1815.

In addition to the packing usually employed in the high-pressure engine, Mr. Trevithick introduces a column or ring of water, which running round the piston renders the whole air-tight. By this means he avoids a great proportion of the usual friction, a very moderate degree of tightness in the packing, being in practice found sufficient to prevent the passage of so dense a fluid as water. The second part of this invention consists in causing steam of a high temperature to spout out against the atmosphere, and by its recoiling force to produce motion in a direction contrary to the issuing stream, similar to the motion produced in a rocket-wheel, or to the recoil of a gun, by which means a rotative action is produced. Mr. Trevithick also describes three other improvements on the high-pressure engine, the latter of which, though only applied to nautical purposes, is by far the most important. It consists in employing a spiral worm or screw similar to the vanes of a smoke-jack, which being made to revolve at the head or stern of the vessel, produces the required motion.

J. T. Dawes, Bromwich, Stafford, Feb. 6, 1816.

The parallel motion usually employed is in this engine rendered unnecessary by the immediate application of the piston to a crank, whose arbor supports the fly-wheel, and communicates motion to the connecting apparatus.

G. F. Muntz, Birmingham, March 2, 1816.

For a method of abating, or nearly destroying smoke, and of obtaining a valuable product therefrom.

A. Rogers, Halifax, March 23, 1816.

For an improved method of setting boilers.

W. Stenson, Coleford, April 9, 1816.

G. Bodley, Exeter, April 27, 1816.

For an engine to work either by steam or water.

J. Neville, Northampton Square, Aug. 14. 1816.

For a new and improved method of generating steam.

W. Losh, Newcastle-upon-Tyne, Sep. 30, 1816.

That part of Mr. Losh's patent which relates to the locomotive engine, consists in an improved mode of connecting and supporting the apparatus by means of pistons working in steam-tight cylinders, which answers the purpose of a spring carriage, and produces a continued equilibrium in the various parts of the engine.

G. Mainwaring, Marsh Place, Lambeth, May 22, 1817.

This invention consists in conveying the steam (after leaving the cylinder in order to be condensed) through a passage or passages surrounded with water supplied from the hot-well. In these passages are fixed a number of metal pipes or tubes which are filled with water from the surrounding casings, and which has its temperature increased by contact with the steam passing to the condenser. The water thus heated is conveyed by a force-pump to the boiler, and a considerable saving of fuel is thus effected.

John Oldham, South Cumberland Street, Dublin, Oct. 10, 1817.

For an improvement in the mode of propelling vessels by the agency of steam.

Moses Poole, Lincoln's Inn Old Square, Dec. 15, 1817.

William Moult, Bedford Square, London, Jan. 15. 1818.

John Scott, Pengo Place, Surry, Jan. 23, 1818.

For an improved mode of propelling steam-boats.

JOHN MUNRO, Finsbury Place, Middlesex, Feb. 12. 1818.

JOSHUA ROUTLEDGE, Bolton-le-Moor, Lancashire, Feb. 27, 1818.

For improvements upon the rotatory engine.

WILLIAM CHURCH, Clifton Street Finsbury Square, April 8. 1818.

THOMAS JONES and CHARLES PLIMLEY, Birmingham, May 7, 1818.

In this specification the patentees describe an apparatus intented to operate either as a blast or steam engine, and the piston is rendered air-tight by means of a column of water.

JOHN MALAM, Marsham Street, Westminster, Aug. 5, 1818.

SIR WILLIAM CONGREVE, Cecil Street, Westminster, Oct. 19, 1818.

The principle upon which elastic vapour is employed in this engine, consists in collecting its force beneath the pressure of a column of water or other heavy fluid, and its effect to produce motion will then be regulated by their re-acting pressure.

To apply this force to the greatest advantage, this ingenious experimentalist recommends the employment of an apparatus resembling the overshot water-wheel, but in this case the wheel is immersed, in a fluid in which it is made to revolve, and the steam entering beneath the hollow boxes or float-boat is made to produce a continuous rotatory motion by its ascent to the surface.

JAMES FRASER, Long-Acre, London, Nov. 12, 1818.

For improvements in the steam-boiler.

Richard Wright, Token-House Yard, London, Nov. 14, 1818.

John Pontifex, Shoe Lane, London, Jan. 7, 1819.

John Seaward, Kent Road, London, April 3, 1819.

For an improved mode of generating steam.

William Brunton, Birmingham, June 29, 1819.

An account of Mr. Brunton's mode of consuming smoke, will be found in a preceding page.

John Oldham, South Cumberland Street, Jan. 15, 1820.

For improvements on a previous patent, dated Oct. 10, 1817.

John Barton, Falcon Square, London, May 15, 1820.

John Hague. Great Pearl Street, Spitalfields, June 3, 1820.

John Wakefield, Ancott's Place, Manchester, June 6, 1820.

For improvements in the construction of furnaces by which a saving of fuel may be effected

William Brunton, Birmingham, 1820.

For an improved mode of constructing furnaces.

Job Rider, Belfast, Ireland, July 20, 1820.

For improvements capable of producing a concentric and revolving eccentric motion, applicable to steam engines, &c.

John Moore, Castle Street, Bristol, Dec. 9, 1820.

For an ingenious, though we fear useless, rotatory engine.

WILLIAM PRITCHARD, Leeds, Yorkshire, Dec. 22, 1820.
For a saving of fuel by the combustion of smoke.

WILLIAM ALDERSEY, Homerton, Middlesex, Feb. 3, 1821.

THOMAS MASTERMAN, Broad Street, Ratcliffe, Feb. 10, 1821
For a rotatory engine which we have already very full described.

ROBERT STEIN, Walcot Place, Lambeth, Feb. 20, 1821.

HENRY PENNECK, Penzance, Cornwall, Feb. 27, 1821.
For lessening the consumption of fuel.

HENRY BROWNE, Derby, March 16, 1821.
For saving fuel and consuming smoke.

AARON MANBY, Horseley, Staffordshire, May 9, 1821.

THOMAS BENNET, Bewdley, Worcestershire, Aug. 4, 1821.

FRANCIS ARETON EGELLS, Britannia Terrace, City Road, Nov. 9, 1821.

CHARLES BRODERIP, London, Dec. 5, 1821.

JULIUS GRIFFITH, Brompton Crescent, Dec. 20, 1821.
For an improved locomotive engine.

APPENDIX (B).

Abstract of Evidence and Reports made by a Select Committee of the House of Commons, on Steam Engines and Furnaces.

MICHAEL ANGELO TAYLOR, ESQ.
In the Chair.

Mr. JOSEPH GREGSON, Surveyor, called in and examined,

Was of opinion that the nuisance that arose from the smoke of steam engine furnaces might be attributed to two causes: one, the putting on the fire or furnace too much crude fuel at one time; the other, from the chimnies being commonly too low, in proportion to the fuel consumed.—Had seen this nuisance effectually removed; but it had generally been attended with an increased consumption of fuel: it was seldom adopted but where the parties had been or were under an indictment.—His own invention consisted in causing all the smoke after it had arisen from the fire, to return into the heat of the fire before it entered into the flue or chimney, and so was consumed; 2dly, By putting on no more fuel at any one time than the smoke of which can be so consumed, and that without opening the furnace door for the purpose; 3dly, By supplying every fire with air, in order to counteract the effect of those winds that operate against the draft.—Had employed it in the fires and boilers of private houses, under steam engine boilers, and in wealding furnaces, where a number of bits and scraps of iron were packed together, and subjected to an intense heat; they were, in that state, then rolled or hammered into one compact body.—The result however in the latter case was, that although every thing acted according to the plans laid down, and the fire was regularly supplied with fuel, and the smoke completely destroyed, yet the heat necessary to weald those scraps of iron together could never be attained, and this was in consequence of the continued repetition of the supply of fuel, damping and preventing that heat coming over which arises after all the volatile parts of the fuel has been driven off; and which heat, being entirely pure, was called a white or wealding heat. The furnaces requiring a white heat and higher degrees, were wealding, melting and smelting furnaces, and vitrifying furnaces, as the making of glass and porcelain. The melting and smelting furnaces were in many instances supplied only with coke, but witness was not aware that a wealding or glass furnace could be at all worked with coke. —Considered that a good effect would be produced by raising the chimneys; as by increasing the draft the smoke would be then more consumed, and by its height more dispersed by the wind.—For every fire consuming one bushel of coals per day, the chimney should be at least thirty feet high, and one foot higher for every additional bushel consumed, measuring from the body of the fire.—At the new steam engine of 100 horse power at the East London Water works, Old Ford, there was a method of consuming the smoke; a singular plan was adopted at the corn mills at Islington, Liverpool; at the corn mills Newcastle, Stafford, a steam engine, of 14 horse power, had worked for nearly a whole day without smoke, owing to the quality of the coal, which was only 5*d.* per cwt.; at the lead mills, Tottenham Court

Road, a small steam engine was worked with coke only; at a brewery in Stafford, a small steam engine, and a large one at the Meddock Mills, Manchester, consumed the smoke on the patent principles; in all, six different engines.—Remarked, however, that under the very best circumstances and contrivances, there were times in which the smoke of crude fuel could not be consumed, viz. at the first lighting of the fire, and at any sudden changes of damping or raising the fire.—The expense of setting up a six horse-engine, on the witness's plan would be about 16*l.*; and a thirty-six one, about 30*l.* or 32*l.*; an old furnace could be altered for 16*l.*, and it would be about one-tenth saving of fuel; it would be upon the gaining, and not the losing side.—In point of fact, the expense of the application of this patent would be shortly saved by the saving of fuel. The same principle would also apply to the steam packets. The smoking of an ordinary chimney was removed by a common fire constructed on that principle. The expense of altering an old engine of a hundred-horse power, upon the new principle, would not be less than 100*l.*—In the making of gas, the coal was only subject to a red heat, and the gaseous vapours of which might be considered as distilled over, while the principal part of the fuel remained as coke; but in a common furnace the coal was entirely consumed, leaving only ashes or a vitrified clinker; the smoke containing much ammoniacal matter, could not be burnt but in a very intense heat, approaching to a white or wealding heat; consequently, when the smoke was burnt under the boiler it was very destructive to the metal; but being burnt upon his principle, that destruction was wholly avoided; the smoke was thus subjected to the required heat, an entire change or decomposition taking place, and the product was principally steam; whereas the coal gas never having been subjected to that heat, there was greater difficulty in freeing it from its impurities; therefore the vapour arising from burnt smoke, was more pure than that arising from the burning of coal gas.

Mr. WILLIAM MOULT, Assistant to Messrs. COOKE, called in; and examined.

The former mode of heating the boilers, was by putting the coals over the bars in the common way; but his improved method was to make the flame come over the coals, which were laid upon an iron plate, and the flames made to pass over the surface of the coals upon the iron plate which lighted the coal at the top, and the red part of the coal was next the bottom of the boiler; by that means, the smoke as it rose from the coals was consumed in its passage over the bars.—The consumption of fuel of the old boiler was regularly eighteen bushels of coals in twenty-four hours; but when altered in this manner, twelve bushels produced a similar effect. The smoke bore no proportion to what it was under the old method. Had put up a small boiler for a steam engine upon the same plan, and found it answer. Thought it would be applicable to soap manufactories, because their boilers were generally long boilers: in some steam engines, it would be difficult to do it, because the fire was obliged to rise in the front, and pass that way.

GEORGE LEMAN TUTHILL, M. D. of Soho Square, called in; and examined.

Believed the atmosphere of London to be prejudicial to health; the accounts which had been published at different times, concerning the relative duration of life in London and in the country, might be considered as having proved that duration to be considerably diminished by a residence in this metropolis. It was probable that this depended upon the atmosphere of London. There was a great variety of causes which contributed to render that atmosphere unfavourable to health; and it might be presumed that the quantity of carbonaceous matter suspended in it, was one of the causes of its insalubrity. The rapid advancement to recovery which we frequently see in sick persons, during a short residence in the country, proved the influence which the atmosphere of London had upon health; there were many diseases incident to the human body, in which the influence of that atmosphere might be more easily detected than in a state of health. In certain diseases of the lungs, especially, it might be proved that the smoke of London was prejudicial.—Conceived that the fog peculiar to London, so different in its sensible properties from any fog in the country, depended upon the smoke of the metropolis, and was prejudicial in many diseased states of the lungs.—The greater the quantity of carbonic acid gas in a given volume of air, the greater would be the insalubrity of that air. But in crowded cities, the air was contaminated from a variety of other causes, which chiefly owed their origin to the exhalations, either from the living animal body, or from decomposing animal and vegetable matter, when the principle of vitality was extinct.—Conceived the smoke arising from steam-engine furnaces might be prevented: it could be effected by making the smoke pass through an ignited tube, whilst the combustion of the soot was there assisted by a fresh current of atmospheric air. Saw no reason why a simple apparatus might not be so contrived, as to render that combustion complete. But it appeared to be deserving of consideration, whether this annihilation of smoke ought to be confined to manufactories. If, instead of burning common coal, that fuel were first divided, as it now was in the gas-light manufactories, into coke and carburetted hydrogen gas, and these were afterwards consumed in union, the brilliancy and warmth which was now enjoyed by the fire-side would, to say the least, be undiminished, whilst the smoke would be entirely destroyed.

This might be tried without any difficulty, by the judicious admission of gas into a common grate filled with coke; the materials in fact would be the same as of our common fires, but employed in a state of greater purity. There was no limit to this mode of destroying smoke: and, should a plan of this nature be hereafter adopted, chimnies, as they were now constructed, would be quite unnecessary; a small tube would be sufficient.

Mr. William Losh, of Newcastle-on-Tyne, called in, and examined.

Considered it impossible to state any thing which could be satisfactory, without referring to a plan. Would only state, that in some engines which had been erected according to witness's plans, the smoke was entirely consumed.—Was of opinion that for smelting ores, long horizontal flues would be advisable, and would nearly do away with the nuisance; but for glass-houses, witness did not know of any practical remedy.

Mr. William Brunton, of Birmingham, Civil Engineer, called in, and examined.

Furnaces for consuming smoke, as they were usually constructed, consisted of two distinct parts: 1st, the grate upon which the coal was consumed; 2dly, the feeding-mouth into which the coal was put (with the shovel) preparatory to its being pushed forward upon the grate, at the end of the feeding-mouth; opposite to that which joined the grate, was fitted a door, in which were holes with covers, for regulating their apertures, by which atmospheric air was admitted at pleasure. The process was thus: whilst the coal already upon the grate was in high combustion, and had ceased to smoke, the coal in the feeding-mouth, being exposed to the heat of the fire, underwent a degree of coking, and the smoke was thereby evolved, which, combined with a portion of air admitted at the openings in the door, passing into the chimney over the hot fire, was consumed. When the fire was to be renewed, the coal thus acted upon, was forced forward upon the grate, still carefully preserving a strong fire of well burnt coal on the farther end of the grate, in order to consume the smoke, which would now be given out by the coal thus brought into active combustion; at this period a much greater portion of air must be admitted, than would be needful when the coal last forced forward had attained its full heat.

The following were the principal objections to the general adoption of this species of furnace: First, the process of coking, or preparing the coal in the feeding-mouth, was very imperfect, and but a small part of the coal necessary to feed the fire was affected by it, so as to give out less smoke when forced upon the grate. Secondly, though the judicious admission of air to enable the smoke to ignite was found advantageous, yet a small excess admitted was found to have a very injurious effect in cooling the boiler; and as the quantity of air required for the combustion of the smoke must vary every moment of the interval between the times of renewing the fire, (perhaps fifteen or twenty minutes,) nothing short of the greatest care and unremitting attention to the admission of the air could accomplish the object with economy. This care on the part of the workmen could very rarely be obtained; and proprietors of steam engines have found that, for want of this, the burning of smoke has been too expensive for them to persevere in.—Witness having turned his attention for some years to this subject, had discovered a method by which the smoke might be consumed economically, and its practicability less objectionable than the methods usually adopted: 1st, by putting the coal upon the grate by small quantities, and at very short intervals, say every two or three seconds; 2dly, by so disposing of the coal upon the grate, that the smoke evolved must pass over that part of the grate upon which the coal was in full combustion, and was thereby consumed; 3dly, as the introduction of the coal was uniform in short spaces of time, the introduction of the air was also uniform, and required no attention from the fireman. As it respected economy, 1st, the coal was put upon the fire by an apparatus driven by the engine, and so contrived, that the quantity of coal was proportioned to the quantity of work the engine was performing, and the quantity of air admitted to consume the smoke was regulated in the same manner; 2dly, the fire door was never opened excepting to clean the fire; the boiler of course was not exposed to that continual irregularity of temperature which was unavoidable in the common furnace, and which was found exceedingly injurious to boilers; 3dly, the only attention required, was to fill the coal receiver every two or three hours, and clean the fire when necessary; 4thly, the coal was more completely consumed than by the common furnace, as all the effect of what was termed stirring up the fire, (by which no inconsiderable quantity of coal was passed into the ash-pit,) was attained without moving the coal upon the grate.—Conceived that in a twenty-horse engine, the increased expense of erection would be between *75l.* and *100l.*

FIRST REPORT.

THE Select Committee appointed to inquire how far it may be practicable to compel Persons using Steam Engines and Furnaces in their different works, to erect them in a manner less prejudicial to Public health and Public comfort; and to report their Observations thereupon to the House;—Have agreed upon the following Report:

That from the advanced period of the Session, at which the appointment of your Committee took place, it was not to be expected that they could form any ultimate decision as to the precise object of their inquiry; but as far as they have hitherto proceeded, they confidently hope that the Nuisance so universally and so justly complained of, may at least be considerably diminished, if not altogether removed.

Your Committee have had under their examination, Men whose minds have been long and practically directed to the extinction of the evil; and from their Evidence, the House will be enabled to judge how far their opinions correspond with those of your Committee.

The disinterested communications made by Persons whose private interests might have led them to a different line of conduct, cannot be too highly valued and extolled.

July 12, 1819.

SECOND REPORT,

From the Select Committee on Steam Engines, Furnaces, &c.

MICHAEL ANGELO TAYLOR, ESQ.
In the Chair.

Mr. Josiah Parkes, of Warwick, Worsted Manufacturer, called in, and examined.

Had practised various methods for reducing the smoke arising from a steam engine, for about six years; had so far been able to accomplish the removal of the nuisance, that for about twelve hours of the day the smoke was nearly invisible; had three boilers, and found that in an hour after lighting the fires, there was heat enough to consume the whole of the smoke arising from them. Had adopted a mode of firing, which was practised with it, greatly conducive to economy of fuel. The daily consumption was twenty-five cwt. to supply the engine, dye-house, and washing coppers; while witness was formerly in the habit of using from thirty-six to forty hundred weight. Employed steam at a pressure of three pounds and a half on the inch, the ordinary pressure of Messrs. Boulton and Watt's engines. Hadapplied the improved apparatus at Messrs. Barclay and Perkins's brewhouse, and was perfectly satisfied with the result; and the destruction of the smoke was nearly complete. But the Newcastle coals made a great deal more smoke and less flame than the Staffordshire, and therefore the destruction of the smoke became a much more difficult object. Had an air-valve to regulate the quantity of air admitted to burn the smoke, which was regulated or closed at pleasure. Had reduced the consumption of coals, by the combined adoption of the mode of firing with the destruction of smoke from thirty-six hundred weight to twenty-five hundred weight, daily. When much smoke was being made, had uniformly found the draft increased, because, by the conversion of the smoke into active flame, the rapidity of its passage was facilitated.—Witness wished to state one further circumstance; that he believed that this plan of burning smoke was applicable to works of various descriptions; and from the effect produced at Messrs. Barclay's brewery, under the three boilers that he had altered, he had received directions from them to proceed with one of the large brewing coppers. Thought it might be applied to annealing and reverberating furnaces, and most close fires. In charging the fire-place, the fuel was gradually pushed backwards as it became ignited, until the whole capacity of the furnace was one mass of coals; the feed-mouth was also then filled up, and the door was closed for the day. This required about two hours. Had stated that the heat was increased by the admission of air to burn the smoke; and to ascertain the amount, had placed a thermometer in such a posture, when the smoke was being consumed by the admission of the air, that it stood at 214 degrees; by the sudden closing of the air-valve, the smoke passing away unconsumed, the thermometer fell to 200°. The appearance of the flame when no air was admitted was red and dusky, intermixed with quantities of smoke; and large quantities of soot were seen to come over from the fire with it; but when the air was admitted, the smoke caught fire with a brilliant white light, and the soot was evidently consumed. It was the combustion of the charcoal which produced the light. The expense of alteration upon witness's plan would be but trifling, probably about 20*l.* or 30*l.* to each boiler; but the premium had not been determined. As a proof of the small expense, and the little time attending the necessary alterations, the three boilers at Messrs. Barclay's brewery were altered in five days.

Mr. WILLIAM PHIPSON, of Birmingham, called in, and examined.

Had paid a great deal of attention to the different plans that had been suggested for burning smoke, and had found them in general, if not always, attended with an increased consumption of fuel; and likewise so much care and skill required from the fireman as was seldom to be found in persons so employed. Had succeeded in considerably diminishing the smoke; but was very far from the perfection which had been obtained by Mr. Parkes. Had been many years acquainted with Mr. Parkes, had watched very closely the improvements that had been making within the last few years at his mill, and had found that he had succeeded for some time past, in a much more complete manner than any other person, and now, considered that he had attained complete success in the consumption of smoke, attended with a considerable saving of fuel. Had seen the furnaces at Messrs. Barclay's brewhouse since they had been altered, and considered the destruction of the smoke as complete there as at Mr. Parkes s at Warwick. Had heard it remarked that there was still a great quantity of smoke at Messrs. Barclay and Co.'s; this arose from three large brewing-coppers, which had not been altered by Mr. Parkes; and the issue of smoke from those coppers was so great as to deceive persons who viewed the works from a distance. Believed Mr. Parkes's plan was applicable to all kinds of boilers, and to most kinds of close fire-places, but not to open fires.

FREDERICK PERKINS, Esq., called in, and examined.

Had been induced to try the method which Mr. Parkes had invented for abating the nuisance of smoke, and had applied this apparatus to two boilers, and believed they were altered in the space of three or four days. When worked there was no smoke excepting at the first lighting of the fire; after the fire was made up, and a regular heat produced, little or no smoke escaped from the chimney when the air valve was open. Believed that the consumption of fuel was decreased; but could not say decidedly.

Mr. JAMES SPURRELL, Brewer, in the Employ of Messrs. Barclay & Co. called in, and examined.

Before the adoption of Mr. Parkes's alteration for consuming the smoke, had made a very considerable portion of smoke from the steam engine, but since the adoption of Mr. Parkes's plan, there was scarcely any smoke after the first of the morning; the fire, when once made up, would last from ten to twelve hours. The application had also been made to another copper, which was called the blowing-off copper, for blowing-off and steaming casks, and there the plan had been as successful as in the steam engine, very much to the satisfaction of the house. Thought that the consumption of coal would be reduced in the blowing-off copper.

Mr. BENJAMIN HAWES, Jun., Soap-boiler, called in, and examined.

Had altered a steam engine furnace under Mr. Parkes's direction, and found a very considerable reduction of smoke in consequence; with respect to fuel, could hardly state whether there is any saving, not having had the returns.

CHARLES MILLS, Esq., a Member of the Committee, examined.

Had been at Warwick, and seen Mr. Parkes's manufactory. Saw the three furnaces in operation, and had an opportunity, by looking into the fire-place, of seeing the difference that was caused in the fire by the opening and shutting of the air-valve. In one case there seemed to be a thick smoke, and the fire clouded; in the other case the smoke was wholly removed, and there was a perfectly clear fire. Went out of the engine-house into the court, to observe the top of the chimney: in one case observed scarcely any smoke visible coming from the chimney; in the other, when the alteration was made, to try the experiment, found a very considerable quantity of smoke to arise.

DUGDALE STRATFORD DUGDALE, Esq., a Member of the Committee, made the following Statement.

Had attended Mr. Parkes's manufactory, for the purpose of seeing the trial of the experiments for consuming smoke by a method he had lately introduced, which appeared to be very simple, and at the same time to be perfectly efficacious.

Michael Angelo Taylor, Esq., Chairman of the Committee, made the following Statement.

I was applied to about six weeks ago to take Warwick in my way to London, for the purpose of seeing the furnaces that were used in the works of Mr. Parkes. I of course appointed a time, and on my arrival at Warwick, in company with Mr. Denman, a member of this house, went into the manufactory of Mr. Parkes, and made every observation I could to satisfy myself of its efficacy in meeting the object which I had in view, that of diminishing the smoke of furnaces erected for the purposes of heating the boilers used for working steam engines, and for furnaces necessary in different branches of manufactures, as well as in brewhouses. I never thought, as far as my present inquiry and observation was directed, that the experiments which had been made, and which had been reported to me by Mr. Parkes and other gentlemen, could lessen the smoke that arises from furnaces used in the smelting of ores, or in the vitrifying of glass. My object in going to Mr. Parkes's was to ascertain how far his experiment could succeed in furnaces of the description I have before alluded to, let their power and extent be what they might, and to judge from every experiment I could there make, whether his plan would be generally applicable. On going into Mr. Parkes's premises I could not perceive the least smoke arising from any chimney in the place, so much so that I was at a loss to ascertain which was the chimney attached to the furnace which supplied the heat for the steam engine. I also noticed very accurately the garden which immediately adjoined the furnace, to see if from the flowers and from the different plants that were in that garden there was upon them the affection of soot or smoke; I could perceive none, though I inspected them very narrowly. I was anxious to make this trial, knowing from experience that the volumes of smoke which issue from the furnaces on every side of the river Thames opposite my own house, actually blacken every flower I have in my own garden at Whitehall. I afterwards went to the bleaching-ground, to judge whether there were any marks of soot or smoke upon the tenter-rail, on which were always hung the different articles for bleaching; I found no appearance of soot or smoke; the bleaching-ground was almost adjoining to the furnace. On going into the place where the furnace was erected, I desired that I might be allowed to assail the furnace in whatever way I chose, and by any experiment I thought proper to make; I mean by the word, assail, to see if by any experiments of mine I could create smoke. In the first place, I desired that the furnace might be opened and raked for a considerable time; no smoke was occasioned, indeed none had issued. I then directed that the valves which secured Mr. Parkes's experiments, and gave effect to his object, should be shut, and immediately the place was involved in smoke, which ceased on resorting to the apparatus again. Not satisfied with that, I desired that a very large body of coal might be put into the middle of the furnace, and that Mr. Parkes's apparatus should remain in the way it did remain when I saw no smoke; I then inspected whether or not, from that immense body of fuel, which I put on adventitiously, any smoke issued; none appeared. On the shutting of the different air-valves, an immense body of smoke issued; this convinced me that his apparatus was perfectly adapted to the object I had long in view; and that though many others had met my observation, yet none appeared to me so simple as his; and I desired Mr. Parkes to come to London, and apply in my name to persons who had steam engines, and see if they would permit him to have their furnaces so altered, that the public might judge whether or no his apparatus would give as fair a promise of success in London as it had done at Warwick. I waited on Messrs. Barclay and

Perkins a few days ago, in company with Lord Harewood, Lord Rosslyn, and several other gentlemen, many of them Members of this House, to see whether the experiment had succeeded in the furnaces attached to that brewery. Every trial was made in my presence, and every trial succeeded; and I have no doubt that if there was occasion, several Members of this House, who were present, would attend and give evidence to the same fact. The success of the experiment was as complete at Messrs. Barclay's as it was at Mr. Parkes's at Warwick."

KIRKMAN FINLAY. Esq. a Member of the Committee, examined.

Had been with other members of this Committee at Messrs. Barclay's brewhouse, and was quite satisfied as to the reduction of the smoke in the furnaces to which Mr. Parkes's apparatus was attached.

MR. WILLIAM BRUNTON, Birmingham, called in, and examined.

Had lately erected eight fire-regulators, all of which had given the greatest satisfaction as regarded burning the smoke, one of which was at the Whitechapel distillery, and another at Liverpool; and they had uniformly effected a saving of coal, which on the average was more than 30 per cent. They were applied to boilers which were originally erected by Messrs. Boulton and Watt, and justly considered upon the best construction of the common furnace. In London, a grate of five feet diameter, consumed three bushels per hour; in Staffordshire and Lancashire, about three hundred weight. First, by the use of the fire-regulator there was much less scoriæ or clinker formed from the same quantity of coal than in the ordinary fire, and that was formed in thin laminæ upon the grate; and in general, while three bushels of coals per hour were consuming upon the grate, the bars were seldom so hot as to discolour writing paper, when pressed against them. Secondly, another effect, of considerable importance in London, was that the fire-bricks which formed a part of the fireplace were not even vitrified, and would therefore last very much longer than in the usual furnace. Thirdly, the boilers, in consequence of the regularity of the heat, might fairly be expected to last much longer, and the supplementary boiler, on which of course the greatest wear would take place, might at any time be taken down, while the principal boiler, with all its connecting pipes, remained unmoved

In the fire-regulator, first, by the very equal distribution of the coal upon the surface of the grate, a thin fire and a sharp draught was maintained, and this was effected by the coal being introduced in small quantities falling upon the whole of the area of the fire in regular succession. Secondly, the coal was introduced upon the fire without opening the fire-door; and this was effected by dropping the coal through the roof of the supplementary boiler. Thirdly, the decomposition of the coal was much more perfect than by the common furnace, and this was effected by the revolving of the grate, which exposed each side of every piece of coal on the grate to the current of the fire passing constantly in one direction across it. Fourthly, the introduction of the coal was completely governed by the steam generated, analogous to a water-wheel, governing by its velocity the quantity of the water permitted to fall upon it; thus, considering the production of the steam as the effect, and the introduction of coal as the cause, the former had a perfect check over the latter, and at no time admitted more coal into combustion than was really necessary for the performance of the work which the engine was then doing. Fifthly, the whole apparatus, being a very simple mechanical arrangement, acted independently of either the skill or the carelessness of the fireman. No coals should be put upon a steam engine fire until they were small enough to pass a three-inch mesh; therefore the necessity of breaking the coal to that size was advantageous. Had lately burned a species of small coal, which had till now been regarded as perfectly useless; and as such there were thousands of tons encumbering the ground in the Staffordshire collieries, incapable of being used with effect in any other furnace than the fire-regulator; and this hitherto supposed rubbish had produced 70 per cent. of the effect of saleable coal; thus bringing into use that which was of no value.—In obtaining the maximum effect thought it advisable to employ a thin fire, with a sharp draft, witness being of opinion that the greater quantity of oxygen brought into contact with the coal in combustion, the greater heat or effect was produced from it. The fire-regulator, and the other modes proposed for burning smoke, stood upon very different grounds; for the value of a boiler as an implement for generating steam, depended upon the quantity of coal which might be burned under it producing a maximum effect; and when any change was made in a furnace, by which either the quantity of coal consumed with a maximum effect was diminished, or the effect of the same quantity of coal decreased, in either case it was injurious. The fire-regulator, while it increased the quantity of coal consumed, increased the effect also, and thereby increased the value of the boiler to which it was attached, as it would raise a greater quantity of steam at a much less expense. The usual plan adopted consisted of a mere change in the form of the furnace, which when appreciated by the proprietors, was of no more value than it was before; but the fire-regulator made a positive addition to the size of the boiler, and was tangible property, and when the proprietor took his stock stood for its own cost.

Mr. James Scott Smith, Distiller, called in, and examined.

Had found that they could consume the smoke to a very great extent, and although it was not completely invisible, yet it was never offensive. Had never understood that it was possible to consume the smoke entirely. The fire-regulator invented by Mr. Brunton had many advantages; 1st. The boilers to which it was attached had their power greatly increased; would last a longer time, and would not be so liable to leak as those on the old plan, which arose from the circumstance of the fire-door not being opened to introduce the fuel, consequently the frequent draughts of cold air were excluded, and the boilers retained an uniform temperature. 2ndly, There was a great saving of fuel, viz. 38 per cent. and this was the average of a three-weeks' experiment with the fire-regulator, compared with the work of three men in a three-weeks' experiment on the old principle.—Considered Mr. Brunton's machinery and alterations altogether, had cost two hundred and fifty pounds.

Mr. William Brunton, again called in and examined.

Conceived that the general expense of altering an old furnace for the purpose of applying improved apparatus, so as to make it effectual to a twenty-horse engine, would be about three hundred pounds; out of which the patentee received about sixty or seventy pounds.

Mr. Peter Whitfield Brancker, Sugar Refiner, called in, and examined.

Had applied to Mr. Brunton to erect an additional boiler, having heard that it would also have the effect of burning the smoke. The apparatus was finished early in May. Had watched the process carefully, and took notice of the difference of the power of the steam, and the quantity of smoke compared with what was emitted formerly, and found that although the apparatus was not perfect for want of the other boiler being also fixed, that the quantity of smoke was very trifling indeed, little or nothing, and that the power of steam was very much increased; the quantity of coals saved, as far could be judged, was something better than one-fourth, about 30 per cent. Was perfectly satisfied with the result, which had caused a good deal of inquiry in Liverpool, many persons having been to examine the apparatus.

Mr. JOHN WAKEFIELD, called in, and examined.

Thirty years ago was acquainted with the town of Manchester. A Mr. Drinkwater had then erected the first mill that was built for mule-spinning in England, in a populous part of Manchester; who did not wish to be offensive to his neighbours there. Witness had turned his attention to that subject with Boulton and Watt's assistance, and consumed a part of the smoke, but it took more coals by ten per cent. than the old mode. In the year 1816 again turned his attention to this subject; it was much wanted in Manchester, and no progress had appeared to be made. In 1817, print works were erected near to Earl Wilton's, at Heaton, on his lordship's property; this was to be on condition that they should make no nuisance on his grounds, which witness had done with a low chimney, not so high as that of the houses: the next was Mr. Jonathan Pollard's, who had an engine set up by Boulton and Watt, and he tried a smoke-burner made by Mr. Robertson of Glasgow, but it did not give satisfaction. His principle was applied, which saved twenty-five per cent. in coal, and consumed the smoke completely, except a little at the renewal of the fire, which was now obviated. Mr. Jowle, a brewer in Salford, was under indictment by the magistrates for a nuisance occasioned by the smoke; he applied to witness, hearing Mr. Pollard's was done, and the alteration was made under his brewing coppers; it had the desired effect, and he so much approved of it, by its saving of twenty-five per cent. and making his premises cleaner, that he had his steam engine boilers altered, for which he was not indicted; and he also had other works where he had alterations made in the same manner.—Had applied the same invention to other works at places for calico-printing, where the pieces lay on the ground, and where the oxides of iron and blacks stained them, which inconvenience was now obviated to their satisfaction. The expense of witness's apparatus for a twenty-horse power engine would be twenty or twenty-five pounds, besides the charge for remuneration.

REPORT.

THE Select Committee appointed to inquire how far it may be practicable to compel persons using Steam Engines and Furnaces in their different works to erect them in a manner less prejudicial to public health and public comfort; and to report their Observations thereupon to the House;—Have, pursuant to the Order of the House, examined the matters to them referred; and have agreed to the following Report:

THE revival of your Committee has afforded a full opportunity of ascertaining how far the reduction of smoke in furnaces of different descriptions can be practically accomplished; and the evidence detailed already, will shew that the object the House had in view has been satisfactorily and effectually obtained.

July 5, 1820.

Stat. 1 *and* 2. *Geo. IV. Cap.* 41. *entitled*

AN ACT

For giving greater Facility in the Prosecution and Abatement of Nuisances arising from Furnaces used in the working of Steam Engines.—To commence Sept. 1st, 1821.

WHEREAS great inconvenience has arisen, and a great degree of injury has been and is now sustained by his Majesty's subjects in various parts of the United Empire, from the improper construction, as well as from the negligent use of Furnaces employed in the working of Engines by Steam: And whereas by law, every such Nuisance, being of a public nature, is abateable as such by indictment; but the expenses attending the prosecution thereof, have deterred parties suffering thereby, from seeking the remedy given by law:—Be it therefore enacted by the King s most excellent Majesty, by and with the advice and consent of the Lords Spiritual and Temporal, and Commons in this present Parliament assembled, and by the authority of the same, That it shall and may be lawful for the court before whom any such indictment shall be tried, in addition to the judgment pronounced by the said court in case of conviction, to award such costs as shall be deemed proper and reasonable to the prosecutor or prosecutors, to be paid by the party or parties so convicted.

And be it further enacted, that if it shall appear to the court before which any such indictment shall be tried, that the grievance may be remedied by altering the construction of the furnace, or any other part of the premises of the party or parties so indicted, it shall be lawful to the court, without the consent of the prosecutor, to make such order touching the premises, as shall be by the said court thought expedient for preventing the nuisance in future, before passing final sentence upon the defendant or defendants so convicted.*

MINUTES OF EVIDENCE,

Before a Select Committee of the House of Commons on Steam Packets.

SIR HENRY PARNELL, BARONET,
In the Chair.

George Henry Freeling, Esq. called in, and examined.

Had the principal management of the Holyhead steam packets.—The Postmasters General having been obliged to purchase all the sailing packets, and to clear the station for the introduction of those vessels, the object was, at first, to make the steam auxiliary to the sailing packets, but it was found that the steam packets could do even more than the sailing packets, consequently two sailing vessels were kept as auxiliary to the steam.—Had three steam packets employed; the Royal Sovereign of 210 tons, and the Meteor of 190; the Sovereign is fitted with two engines of forty-horse power each, and the Meteor with two engines of thirty-horse power; they were both constructed by Boulton and Watt, and the vessels built in the river Thames, by a person of the name of Evans, at Rotherhithe, on purpose for the service, under the inspection of the officers of the Navy Board; they were built upon Sir Robert Sepping's principle of the diagonal fastening, and made particularly strong. The third is the Ivanhoe, of 165 tons; it was formerly on the Holyhead station as a private vessel, and has an engine of fifty-six horse

* By Section 3. it is provided, that the provisions of this act shall not extend to Steam Engines employed solely in the working of mines, or smelting of metals.

power. The general effect of the experiment, in regard to maintaining a communication between the two countries, has been, that the intercourse has been very much facilitated; it is now almost reduced to a certainty. In the year preceding the introduction of the steam vessels, a hundred mails exactly arrived in London after they were due, and in the nine months that the steam vessels have been running since May last, there have been twenty-two only. The weather at the beginning of the winter, was worse than has been known for more than sixty years. Had proof that the steam packets would go to sea in weather when sailing packets could not have gone to sea; the captains had always considered that it would not be prudent to go to sea, if they were obliged to be under a three-reefed mainsail, and the steam packets had gone out in weather in which the sailing packets would have been obliged to be so. The average of the passages of the Sovereign from Howth to Holyhead, was six hours and fifty-seven minutes, and the Meteor seven hours and four minutes and a fraction. To Howth, the Sovereign seven hours, thirty-six minutes and a quarter; the Meteor eight hours and thirteen minutes: the shortest passage was from Howth, five hours and thirty minutes. The best point for a steam vessel, in very bad weather, was directly head to wind; both wheels could then act at the same time. The captains sometimes kept the vessel away, when it was blowing very strong, two or three points; then, when they got on the opposite coast, they would take in their sails, and steam to the harbour in smoother water. Conceived that the success of these two vessels, the Sovereign and the Meteor, might be attributed to the superior manner in which they were constructed. Had attempted to gain some information about every steam vessel which had been built, and was convinced those vessels would do what no other vessel could do; they would go to sea in weather when nothing else could.

Attributed it not only to the machinery, but to the weight of the hull; a lighter vessel in a heavy sea would be checked, but those vessels had from their weight a momentum so great, that it carried them on when a lighter vessel would have been checked; the weight acting as a fly-wheel.—Was of opinion that three packets were a sufficient number for maintaining the communication between Holyhead and Dublin, so that two should sail every day. With the view that there might be a sufficient time allowed for looking over the machinery and the vessels, it was arranged that they should each be six days at sea and three days in harbour, which afforded ample time for inspecting the machinery; and that had been fixed in a great measure with reference to the engineers themselves, who stated that that time was more than sufficient for it.—There had been some accidents to the engines, but these had been attributed to the use of cast iron; the cross bars and the beams were of cast iron, and if any water was in the cylinder at starting, the check caused the cast iron to break; had now got them made of wrought iron, but the lower beams of the engines were still made of cast iron; there must be some part of the engine left to give way in case of any emergency, which was better than destroying the cylinder.—Believed the Postmasters General had some idea of trying whether what are called Scotch engines, might not be better for a third vessel, in case of one being built; they were more simple, though perhaps not quite so efficient, not so liable to derangement, and were consequently better for a heavy sea; and if the vessel was properly built, witness did not think there could be any great difference in the rate of speed.—The boilers in the Holyhead packets were low pressure. Believed Mr. Watt was the inventor of the original high-pressure engine, but afterwards abandoned it on account of the danger.—No cases of late had happened of accident from the bursting of boilers.

GEORGE HENRY FREELING, ESQ. again called in, and further examined.

Wished to explain some parts of his evidence given in a former day. Did not put any fuel or coals over the boiler, which was the cause of the Robert Bruce catching fire and being burnt. The coals are stowed in iron cases made for the purpose, in the engine room. The other point was as to the Ivanhoe. Witness was asked whether she was so strong as the other vessels, the answer was simply "No." But she was not three years old; she was inspected at Liverpool a short time ago, and appeared as strong as any of the steam vessels, except those on the Holyhead station.—On board the Royal Sovereign there are twenty births, and two rooms, one for ladies and one for gentlemen.

CAPTAIN WILLIAM ROGERS, called in, and examined.

Commanded one of the Holyhead packets. Had crossed in the Meteor on the 5th of February, in the heaviest sea witness had seen during the eight years he had been in the station. Went in the Meteor on the 5th of February, when no sailing packet could carry canvass; they must have laid to; left at nine at night, and arrived at six the next morning. Was satisfied his steam vessel was capable of performing what no sailing vessel could do. Had found that a steam vessel was capable of making her passage much sooner, under all circumstances, than a sailing vessel; in one-half of the time upon the average. With the wind at W. N. W. blowing hard, and leaving Holyhead in a gale of wind, witness had found a steam vessel had been much easier than a sailing vessel; their extreme length overcame the short sea.—In building a steam boat she ought to have a fine entrance and her bow to flear off, not to shove any water before her; any water she shoves before her must be an impediment to the sailing; she should have a fine entrance, a good line of bearing, and her transome pretty square, and not too high; the more a vessel was stopped from pitching and rolling, the quicker she would go. Had found with regard to the Scotch boats that all their transomes were too high and too narrow, the consequence of which was, that with a head sea they would go with their stern under. Had seen them go boat and every thing under; the transome being square and low and fine under, so as to give them a right line of bearing, would stop their pitching and rolling, and make them easy on the sea, and add to their speed. The Meteor and the Sovereign were filled up solid to the floor-head, caulked inside and out, having no tree nails, but bolted and copper nailed. The bolts were driven upon a ring, and clenched at both ends. The diagonal fastening is a plank three inches thick, fore and aft, three and a half thick midships, and nine wide, leading from the floor-head to the shelf, taking in five or six timbers; and filled by truss pieces into triangles, so that it was almost impossible that the form of the vessel would alter. Would prefer Boulton and Watt's engine to any other; their boilers were very superior, and never short of steam. Boulton and Watt had been accustomed to vessels for rivers, and the engines were made rather too slight for the channel; the shaft was hollow, and of cast iron, but they had been replaced by solid shafts. Sails assisted the vessel very much; had used them every way, except going head to wind, within four points of the wind. Had found the Sovereign go as fast in a calm as at any other time. It must not be thought that a steam boat running before the wind in a gale and a heavy sea, ought to make the quickest passage, as they were then obliged to shut off half the steam, or great part of it; for should the full power be on, the wheels running two or three times round without touching any thing between the trough of the sea, and then being brought up all at once, something would probably give way. Was of opinion, that in the event of the engine failing, with the assistance of sails and the anchor, the packet might be kept in perfect safety.

CAPTAIN WILLIAM ROGERS, again called in, and further examined.

On the 16th of May, blowing hard from the S. W. 3. P. M. witness left Gravesend on board the Sovereign, in company with the Meteor steam packet, with seven or eight men on board of each. At nine anchored in the Downs, blowing very hard; she rode very easy with thirty five fathom cable in five fathoms water. On Thursday, the 18th, fresh gales from W. S. W. 5. A. M. weighed anchor and steamed for Portsmouth; wind dead on end. 4. P. M. made the Owers Light; hazy weather. 9. P. M. very heavy gales and thick weather. Asked the engineer what coals he had on board, and was told five hours; were then obliged to steam in for the land. At 10 made the Nab Light upon the starboard quarter, about a quarter of a mile; shortened steam for the Meteor to come up. At 11 anchored at St. Helen's, in six fathoms water, with forty fathom cable; hard gale. On the 19th, 3. 30. A. M. weighed anchor and ran into Portsmouth harbour; made fast to one of the buoys alongside His Majesty's ship the Queen Charlotte; were then employed in getting coals in, and very bad they were. At 4. 20. got under weigh and steamed out of the harbour, and at 9 passed through the Needles against a flood tide. This day was fine weather, and light airs from the N. W. At 1. 30. P. M. anchored in Falmouth harbour, and remained a week to clean the boilers, to caulk the decks, and so on. Sailed from Falmouth the 26th, wind N. W. fresh gales. At 5. 30. past the Longships Lights. At 9 in a squall, with heavy rain, the wind shifted to the N. N. E. blowing very heavy and a heavy sea, the vessels going from three knots to three and a half, head to wind, blowing hard. At 7 made Lundy Island, bearing E. by S. passed several vessels lying to ; passed a large smack, lying to, under close reefed mainsail. At 8 made sail upon the vessels ; stood more to the southward into the Bristol Channel, to smoothen the water. At noon more moderate ; water smoother ; down all sails and steamed for Milford. At 5 anchored in Milford, found several vessels had been out in the gale and obliged to put back ; the vessels that had been put back, bound to Liverpool, said they had never experienced worse weather before for many years. On Sunday evening, the wind more moderate, and from N. to N. E. At 8 P. M. got under weigh. On Monday, at 4 P. M. arrived at Holyhead. Had been five days performing the voyage, with the wind right a-head down the English Channel and up the Irish.—Considered it impossible for any square-rigged vessel, from a first-rate down to a sloop of war, to have effected the same. In the Downs passed several Indiamen, and 150 sail there that could not move down Channel, and at the back of Dungeness passed 120 more. Witness would describe to the Committee the exact improvement he would recommend as to the construction of a new steam vessel.—Should make her a foot narrower, and raise her floor-heads a little, take off the roundings, with her engines put nearer the centre, the boilers much lower, and the wheels narrower. Had observed in vessels with wide wheels, the lee wheel was a great deal under the water, and the other out; by the width of them it increased the angle ; and although a wide wheel was of great advantage in a river, it was a great disadvantage in a sea : supposing there were two forty-horse power engines, would not have more than a seven-feet wheel; and if there were two thirty-horse power, six and a half would be sufficient. On board the Meteor and Sovereign, to prevent accidents from fire, there is water all round the furnaces and boilers, and they are kept three feet from the bottom, about fifteen inches from the side, and twelve inches from the deck; and it was quite impossible that any accident could happen; there is water under and on the sides of the fire places: in those boats that are fitted up upon the Clyde there is nothing but brick and mortar below the furnaces. The Meteor and Sovereign have iron.

CAPTAIN WILLIAM ROGERS, again called in, and further examined.

In the event of building a new steam packet, would recommend that she should be built on Sir Robert Sepping's plan, as to mode of fastening, &c. only a little finer at each end, and one foot narrower than the Sovereign and Meteor, to be 95 feet in the keel, 105 or 106 upon deck, and 19 feet in the beam, about 180 tons, her transomes square, and not very high out of the water. The improvements in the engine would be to make them a little stronger, and the boilers a little longer, keeping them more from the side of the vessel, so that the heat might not affect her, and more room to go round them, and to put the boilers lower down. It would help to prevent their rolling. The Talbot's boilers are a little higher than the Ivanhoe's; by putting the Ivanhoe's lower down, found she did not roll near as much. Would have the two main beams put close to the wheel, which would reduce the weight very much, and strengthen and make the vessel much easier, as by being so far asunder adds great weight, and acts as a lever. There sho ld be twelve paddles, about seven feet long, and nineteen to twenty-one inches wide. The engine should be something between sixty and eighty-horse power, but this must depend upon the fineness of the vessel, and the water she will draw. If the engine was to be made much stronger, in that case must not go further than two thirty-horse power engines, as then the weight might be too much. Should recommend, with regard to the sails, a large lug forward, and a jib, and a fore and aft mainsail; and in case any thing should happen to the engine, would keep a square topsail on board, and a gaff topsail aft, but not to be used except in case of necessity.

MR. JAMES BROWN, called in, and examined.

Belonged to the house of Boulton and Watt. Superintended the erection of the engines on board the Meteor and Sovereign steam packets. Generally considered the working parts, the cross bars, the side rods, and the side beams most liable to fail. Attributed the cross bars having broken on board these vessels, in some instances, to want of caution on the part of the engineer, and in other cases from stress of weather; they were now made of wrought iron for sea vessels. Had put engines into most of the vessels on the river Thames, the Dover station, and Leith. Upon the Leith station one or two cross bars had broken, and that was entirely from want of caution on the part of the engineer in starting the engine; the other vessels were the Dasher and Arrow, on the Dover station; there was one accident of a cross bar breaking in the Dasher. Did not conceive that an engine could be made without being liable to break; there were some parts so small that they must give way, and it was better the most insignificant parts should fail, than some of the principal ones, because they were easier repaired. Would require more room than was granted in the Sovereign and Meteor, as it would be better to get round the boilers entirely; they could then be painted every month, which would be a great advantage, for the action of the salt water is very detrimental to the iron; they ought to be painted every three weeks, or every month; where a boiler will not leak with fresh water, it will with salt water; and it forms an incrustation upon the surface by exposure to the atmosphere; they should be pumped out, and frequently cleaned. Should think the engineer generally looked over his engine every morning before he started, to see that every part was properly oiled, and that every screw and joint was tight. If so inspected, once a fortnight would be sufficient to examine the packings of the slides and pistons. Would make the paddles for a new engine the same as for the Leith vessels; they are made with wooden floats, in one piece, having three sets of arms, and the bolts are of a peculiar description, which allows the paddles to slip from the outer end of the arm towards the centre of the shaft, by which means a vessel may use her sails the same as any other vessel; should any thing give way, there would be nothing but the arms in the water. That sort of excessive violence which may contribute to injure and break the engine, may be occasioned by any of the working parts getting loose, any of the screws getting between the finer parts, or the water getting upon the top of the piston in starting the engine, the latter of which is occasioned by the condensed steam that forms from the boilers after the passage is over; it gets through the steam pipe to the cylinder and condenses there, perhaps to the depth of 18 inches on the top of the piston, and if the engine is started suddenly without it being cleared away, it has not time to get through the thoroughfares. The consequence is, that it is jammed between the top of the piston and the under side of the cylinder cover, and risks the breakage of some part; as water will not compress, something must give way, and the weaker parts of course will go first. Hardly thought it possible that fire could take place without great negligence, because the furnaces were completely surrounded with five inches of water round every part, and it was only in raking out the fire, and neglecting to water it, that any accident could take place ; it was raked out upon an iron floor. No rolling of the vessel could throw the fire about, she must be pitching to a great degree if that was to take place. The coals are carried in boxes in front of the boilers, so as to be right and left for the fireman. Hardly conceived on board these vessels, where condensing engines were used, that it was possible that a boiler could burst; they were generally provided with two safety-valves, and the steam used was about from two and a half to three pounds pressure upon a square inch, at which pressure it blew off by the safety-valve of its own accord.—Boulton and Watt had made their safety-valves for many years in the way they now are, inaccessible for any person to load them by putting additional weight upon them ; had seen the Scotch engine men, in starting their engines place their feet upon the safety-valve. Supposing that the safety-valves should get choked, the steam would come off at the feed pipes ; it would not give way under any circumstance, not even though the valves were choked, the pressure was so extremely small : the boilers were calculated to sustain 50 times the pressure required of them.—If any part of the boiler, by length of use, became very thin, and gave way, it would merely rend, if malleable iron. The accidents that happened from boilers, sometime ago, arose from their being made on the high-pressure principle, and being made of cast iron. The Meteor consumed about seven bushels of coals per hour, rather under; she was then working above her full speed ; the Sovereign was from nine and a quarter to nine and a half ; the Meteor had two thirty-horse engines, and the Sovereign two forty-horse engines ; the latter, when using that quantity of coal, was going about nine miles and three-quarters per hour. A bushel of good Newcastle coal was reckoned equal to one hundred weight of Scotch coal ; so that it came to very nearly the same thing ; the Scotch coal generally burnt very free, and so did the Staffordshire, but the bushel of Newcastle coal was equal to a hundred weight of either. The standard bushel of Newcastle coals should weigh eighty-eight pounds ; the best Wall's End, eighty , and the Wylam, seventy-seven. Had found, that by actual weight, the specific gravity of the Wylam coal was much under that of the Wall's End ; the latter was not good for working engines. The best coal for steam engines was the Halbeath or Inverkeithing, from a place in Fife called Inverkeithing ; its peculiar value lay in burning free, and becoming a complete white ash, without caking upon the fire bars ; the sulphur in coals would destroy the fire bars in a short time. Had found inconvenience from salt forming in the boilers ; when the man-hole cover was taken off, and they were exposed to the atmosphere, the water then became crystalized, which rendered them very difficult to clean ; this had been avoided by constructing pumps or cocks to let the water through the side of the ship without the man-hole being opened. The salt water oxidates iron very rapidly indeed, if it be allowed to lie upon it. On board a steam vessel a boiler constantly in use would last from four to five years with care ; it was the only part of the machinery subject to decay. These engines, upon the whole, require a considerable degree of care and superintendence, and skilful engineers ; more care than the land engines ; every engine requires great care.

CAPTAIN JOHN PERCY, called in, and examined.

Commanded the Hero steam packet, from London to Margate ; and the Victory, belonging to the same company ; commanded the Victory for three years ; the Hero was built last year. The Hero has two engines of fifty-horse power, made by Murray and Fenton, of Leeds, and carries 427 tons. The Hero consumes pretty well three quarters of a chaldron, or 27 bushels of coals, London measure, per hour ; in general, they make away with six chaldrons in the passage, that was owing to the want of flues ; had not flues enough ; had four furnaces. The distance from London to Margate was about eighty-four miles : generally made the passage in about seven hours and a half, that was the average passage ; one passage was made in six hours and sixteen minutes, with the wind and tide. The paddles were eight feet in the centre. Had been trying an experiment with twelve paddles on a wheel, and it answered very well ; last summer worked with sixteen paddles, three feet and a half between each. The paddle of the Victory was five feet and a half, in one place ; the Hero worked more with the paddles being further apart, and they were lowered a little. They take of the water about seventeen or eighteen inches, and the engine makes thirty strokes per minute. She once did up to thirty-one, but twenty-nine and thirty is about the average.—The passage from London to Margate required, on an average, about seven hours and a half, and they went at the rate of between eleven and twelve miles per hour.

REPORT.

THE Select Committee appointed to inquire into the state of the roads from London to Holyhead, and from Chester to Holyhead; into the regulations for conveying his Majesty's Mail between London and Dublin, and between the Northern Parts of England and Dublin, and between Dublin and the interior of Ireland ; and into the state of the Mail Coach Roads, in Ireland ; and to report their observations thereupon; together with the MINUTES of the EVIDENCE taken before them, from time to time, to the House; —Have, pursuant to the order of the House, further examined the Matters to them referred, and have agreed to the following REPORT:

YOUR Committee have proceeded, in compliance with that part of the instructions of the House, which relates to the conveyance of his Majesty's Mails between Holyhead and Howth, to examine into the circumstances attending the establishing of Steam Packets, at Holyhead, in the course of last year. For this purpose two vessels, called the Royal Sovereign and Meteor, were built, by order of the Postmasters General, in the River Thames, on a plan to give to them the greatest possible strength, and the advantage of the most improved engines. The Evidence which has been given to your Committee by a Commander of one of them, Captain Rogers, leaves no doubt of the practicability of performing the Post-office service at Holyhead, by Steam Vessels, with as great safety as it can be performed by Sailing Vessels, even in the most tempestuous weather; and at the same time by voyages, on an average not exceeding one-half of the number of hours which formerly was the average of the voyages of the Sailing Packets. But your Committee are not as yet prepared to enter into all the details of this important subject; their object in presenting this Report to the House, is merely to convey to the House an opinion they have come to, in consequence of the evidence of Mr. George Henry Freeling, and of Captain Rogers, that the Postmasters General ought immediately to give orders for building a new Steam Packet, so that at least there should be three Packets on the Holyhead Station, before the commencement of the next winter, of that peculiar construction which has enabled the Sovereign and Meteor to go to sea throughout the whole of the last winter.

Your Committee strongly recommend the same general plan of construction should be adopted in building a new Packet, as that on which the Sovereign and Meteor were built; and also, that the Engine should be made by Messrs. Boulton and Watt. They also recommend that the suggestions of Captain Rogers should be attended to in all matters respecting the building of a new Packet, as those suggestions will come from a person who appears to your Committee to possess great knowledge in seamanship and ship-building, and by the experience of commanding a steam vessel through a most tempestuous winter, to have made himself master of the best method of managing one at sea, and also of all the main properties of the mechanism of the engine.

Your Committee have annexed to this Report the evidence of Mr. George Henry Freeling, Captain Rogers, Mr. J. Brown, and Captain John Percy, and also certain Queries which they have sent to several persons who have had the most experience in constructing and navigating steam vessels. They intend to continue their inquiries upon this interesting subject, and hope to present to the House a full Report upon all its details before the close of the Session.

April 2, 1822.

AN ACT

To ascertain the Tonnage of Vessels propelled by Steam.

59th *Geo.* 3. *Cap.* 5. Be it therefore enacted, &c. that the rule for admeasuring ships or vessels to be propelled by steam, shall be as follows; that is to say, the length shall be taken on a straight line along the rabbet of the keel of the ship, from the back of the main stern post to a perpendicular line from the forepart of the main stem under the bowsprit, from which, deducting the length of the engine-room, and subtracting three-fifths of the breadth, the remainder shall be esteemed the just length of the keel to find the tonnage. And the breadth shall be taken from the outside of the outside plank in the broadest place of the ship or vessel, be it either above or below the main wales, exclusive of all manner of doubling planks that may be wrought upon the sides of the ship or vessel; then multiplying the length of the keel by the breadth so taken, and that product by half the breadth, and dividing the whole by ninety-four, the quotient shall be deemed the true contents of the tonnage, according to which rule the tonnage of all such ships and vessels shall be measured and ascertained; any thing in any act or acts to the contrary notwithstanding; provided always, that it shall not be lawful to stow or place any goods (fuel for the voyage excepted) in the said engine-room; and if any goods shall be so stowed or placed, such ship or vessel shall from thenceforth be deemed and taken to be a ship or vessel which has not been admeasured according to the rules of this Act, and liable to all the consequences thereof.

APPENDIX (C)

Chronological Catalogue of Works descriptive of the Steam Engine.

BRANCAS. Le Machine, folio. *Roma,* 1629

Marquis of Worcester's Century of Inventions, 12mo. *London,* 1663, 1746; *Glasgow,* 1767; *London* 1786, 1813

Papin. Recueil de Pieces, 8vo. *Cassel,* 1695

Savery. The Miner's Friend, 8vo. *London,* 1702

Isaac de Caus. New Invention of Water Works.* *London,* 1704

Ars nova ad Aquam Ignis adminiculo efficacissime elevandum. *Cassel,* 1707

John Allen. Narrative of several New Inventions and Experiments, particularly the navigating a Ship in a Calm, and Improvements on the Engine to raise Water by Fire, 8vo. *London,* 1730

Voyage de La Motraye, en Europe, Asie, et Afrique, folio, 3 vol. (See vol. iii. p. 360.) *La Haye,* 1732

Hull's Description of a new invented Machine for carrying Vessels or Ships out of, or into any Harbour, Port or River, against Wind and Tide, 12mo. *London,* 1737

---

* This tract contains a very accurate account of Savery's Engine, with Plates.

Desaguliers's Course of Experimental Philosophy, 4to. *London*, 1763

Blackey sur les Pumpes à Feu, 4to. *Amsterdam*, 1774

Falck's Description of an Improved Steam Engine, 8vo. *London*, 1776

Leupold Theatrum Machinarum Generale, folio, *Lips*. 1780

Belidor. Architecture Hydraulique, 4to. *Paris*, 1782-90

Bossut. Traité Théorique at Expérimental d' Hydrodynamique, 8vo. 2 vol. *Paris*, 1786-7

Senphin de Mon Copi, Œuvres de. 1787

Prony. Nouvelle Architecture Hydraulique, 4to. *Paris*, 1790-6

Boulton and Watt's Directions for Erecting their new invented Steam Engine, 8vo.

Short Statement of Boulton and Watt, in Opposition to Hornblower's Renewal of Patent, 8vo. *London*, 1792

Langsdorf. Lehrbuch der Hydraulik, 4to. *Altenb*. 1794

Smeaton's Reports, 4to. *London*, 1797

Curr's Coal Viewer, and Engine Builder, 4to. *Sheffield*, 1797

Walker's System of Philosophy, 4to. 1799

Nieuwe Verhandelingen van het Batassch Genootschap, *Rotterdam*, 1800

Walker on Draining Land by the Steam Engine, 8vo. *London*, 1813

Buchanan on Propelling Vessels by Steam, 8vo. *Glasgow*, 1816

Dodd on Steam Packets, 8vo. *London*, 1818

Rees's Cyclopædia, *Art*. Steam Engine.

Robison's Mechanical Philosophy, 8vo. *Edinburgh*, 1822

PHILOSOPHICAL TRANSACTIONS.

Papin's Engine, vol. xix. p. 481. 1697

Savery's Steam Engine, vol. xxi. p. 228. 1699

Payne's new Invention of expanding Fluids, vol. xli. p. 821 1741

Blake on Steam Engine Cylinders, vol. xlvii. p. 197. 1751

Smeaton on de Moura's Improvements on Savery's Engine, vol. xlvii. p. 436. 1751

Fitzgerald on increasing Steam by Ventilation, vol. l. 53, 370. 1757

HISTOIRE DE L'ACADEMIE ROYALE DES SCIENCES.

Amontons' Fire Wheel, p. 101. 1699

Dupuy's Steam Engine, with De Moura's Improvements, p. 111. 1740

Gensanne's Engine, p. 60. 1744

Lavoisier on the Expense of Steam Engines, p. 63. 1771

MACHINES ET INVENTIONS APPROUVEES PAR L' ACADEMIE.

Mey and Meyer's Engine, vol. iv. p. 185. 1720

Bosfraud's Steam Engine, vol. iv. p. 191, 199. 1720

Gensanne's Engine, vol. vii. p. 222. 1734

INDEX.

THE END.

J. and T. Bartlett, Printers, Oxford.

Savery's Engine, improved

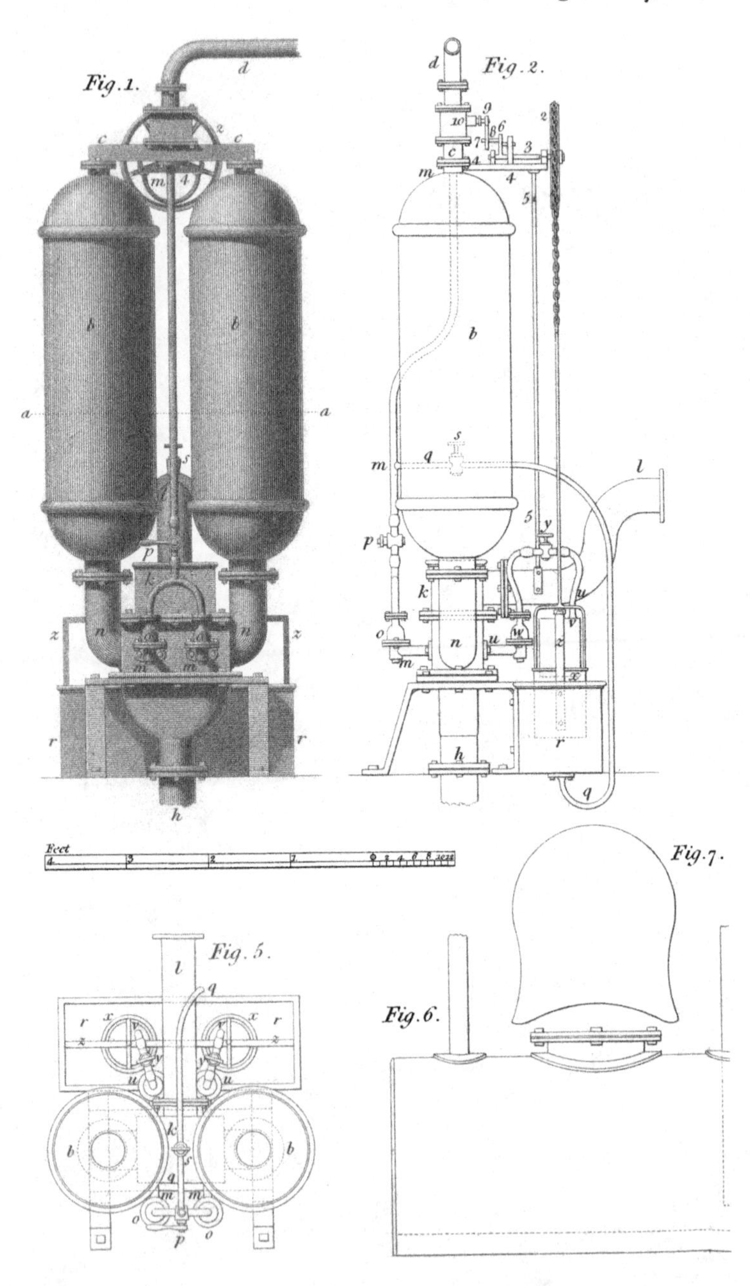

Drawn by J. Clement.

London, Published by J. Taylor, at the Architectural

by J. Pontifex.

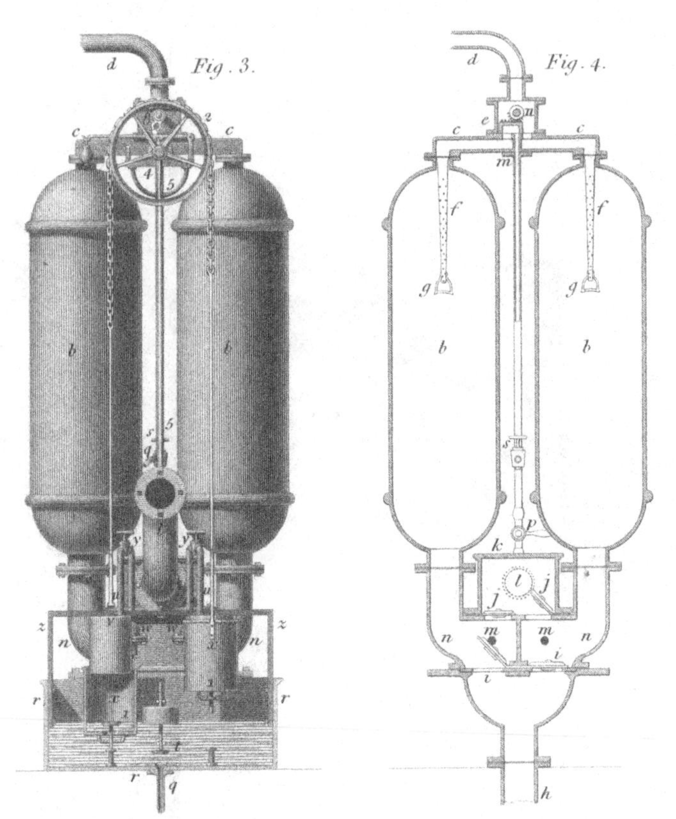

Brancas's Engine.

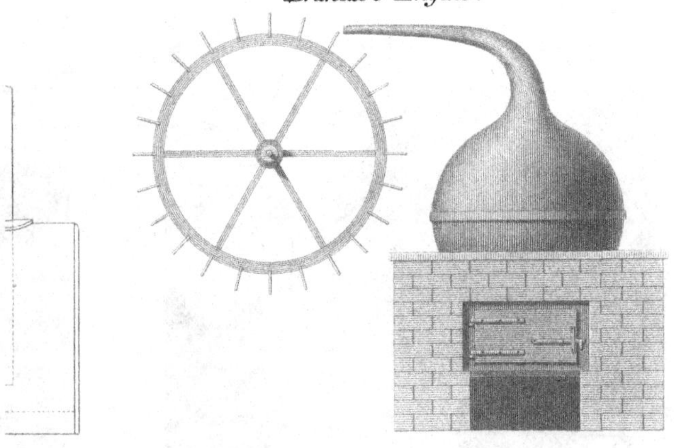

Engraved by G. Gladwin.

Library. 59 High Holborn. June 1822

Atmospheric

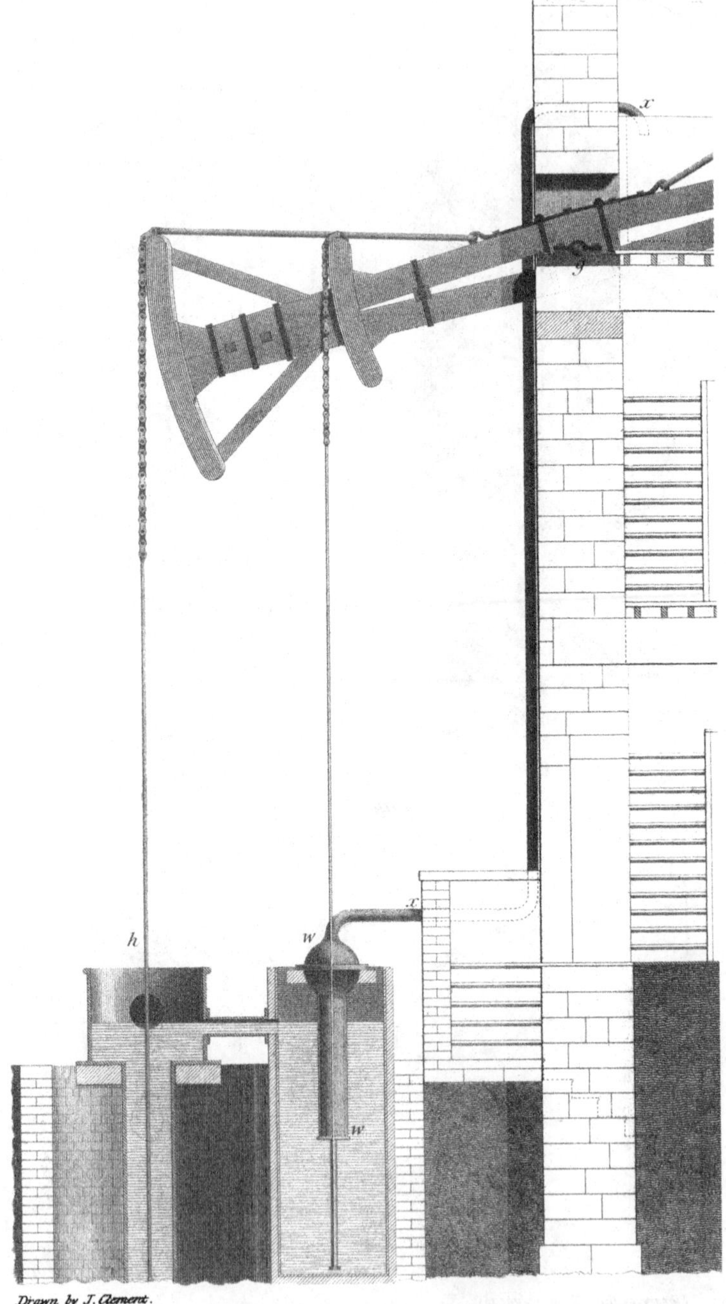

Drawn by J. Clement.

Steam Engine.

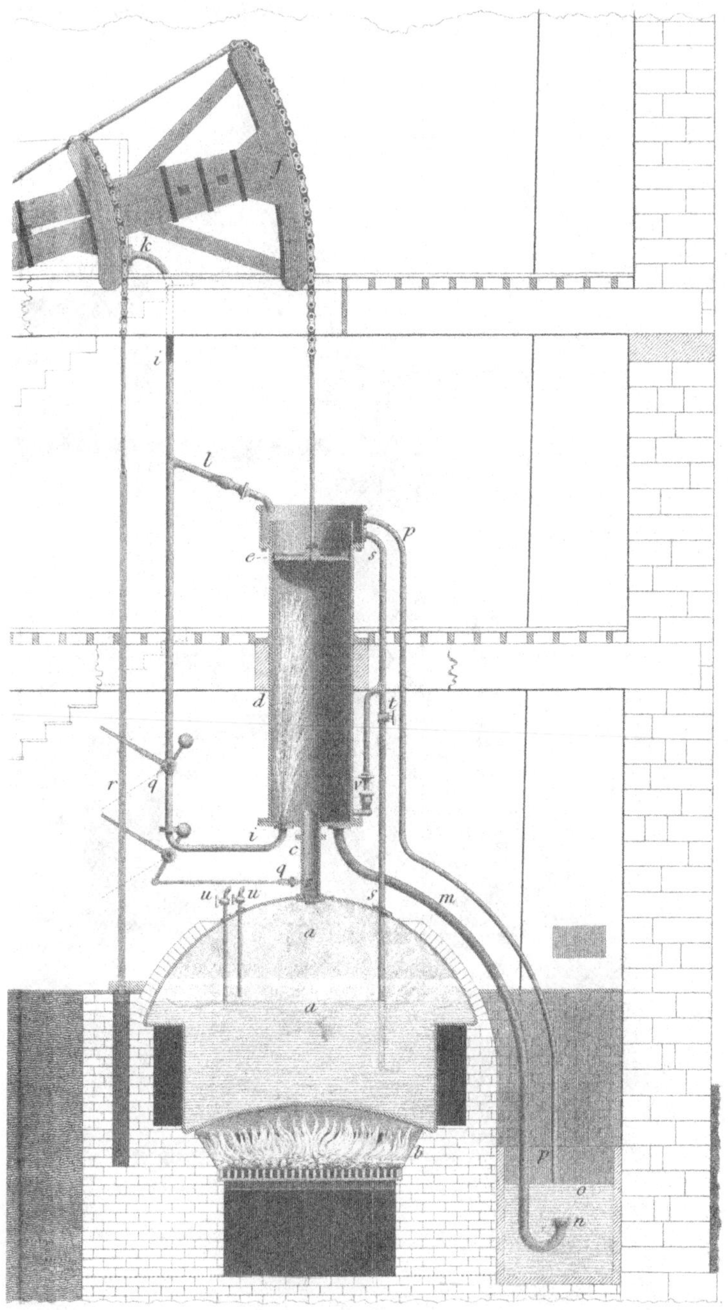

Eng^d. by Edwd. Kennion.

Single Acting Engine

Drawn by J Clement.

Plate. III.

by Boulton and Watt.

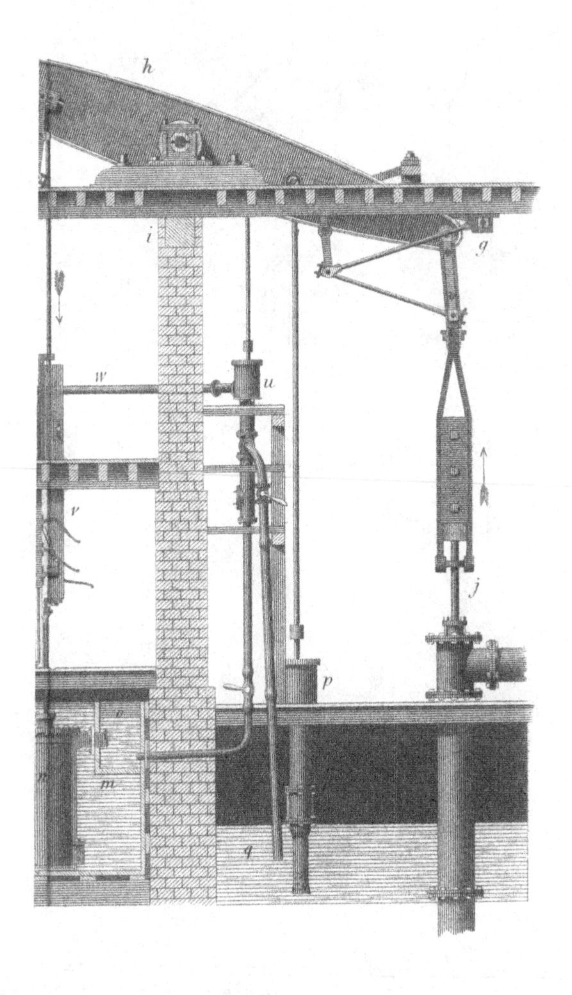

Engraved by G. Gladwin.

Section of Mefs.rs Murray & Woods Steam pipes and Valves.

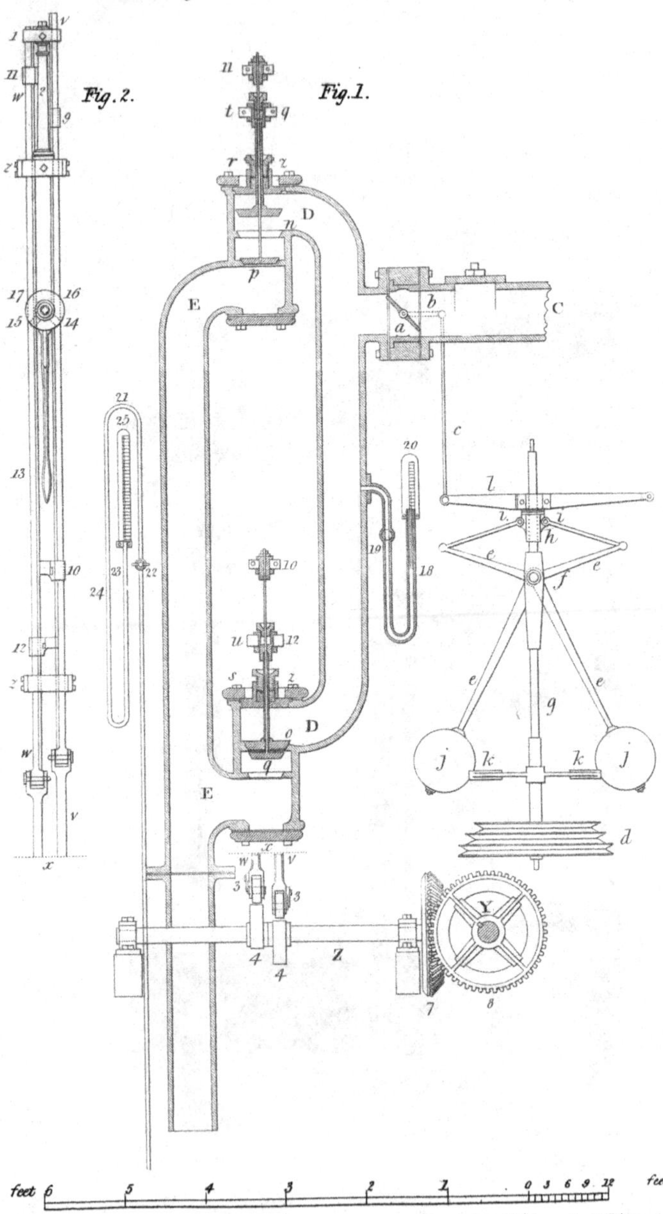

Drawn by J. Clement

London. Published by J. Taylor at the Architectural

Plate V.

Section of the Cylinder and working Valves to Maudslays Engine.

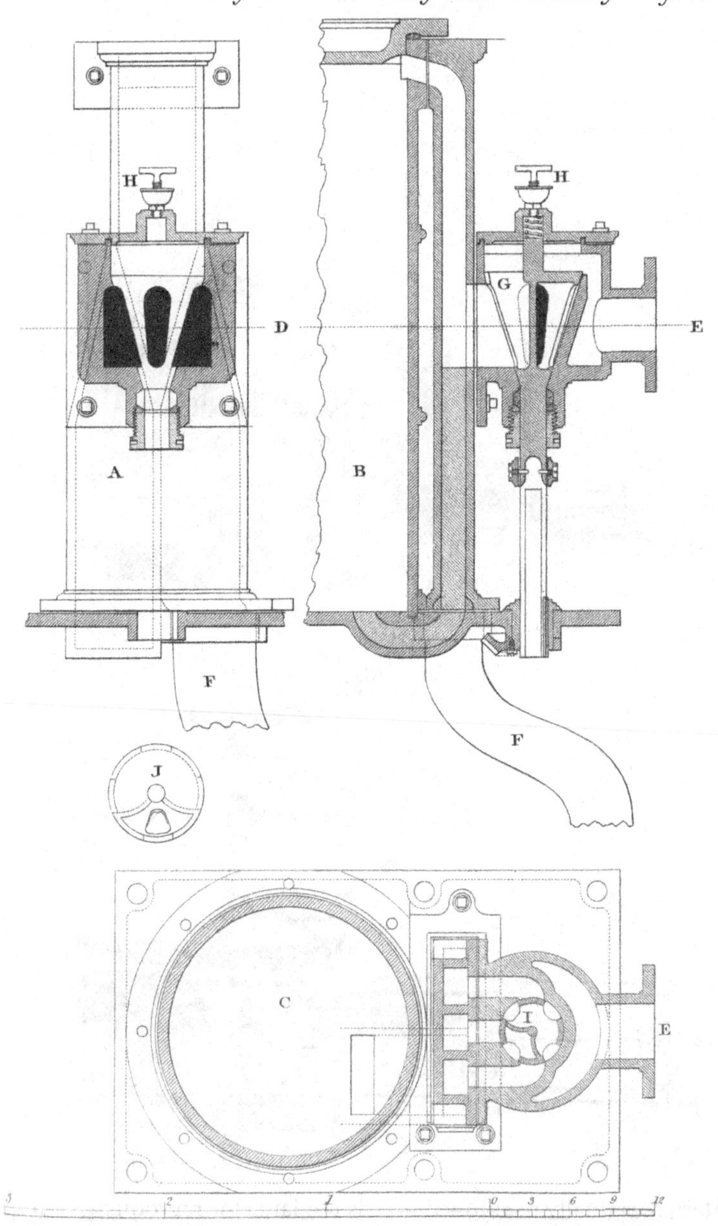

Library, 59 High Holborn, June 1822

Engraved by G. Gladwin.

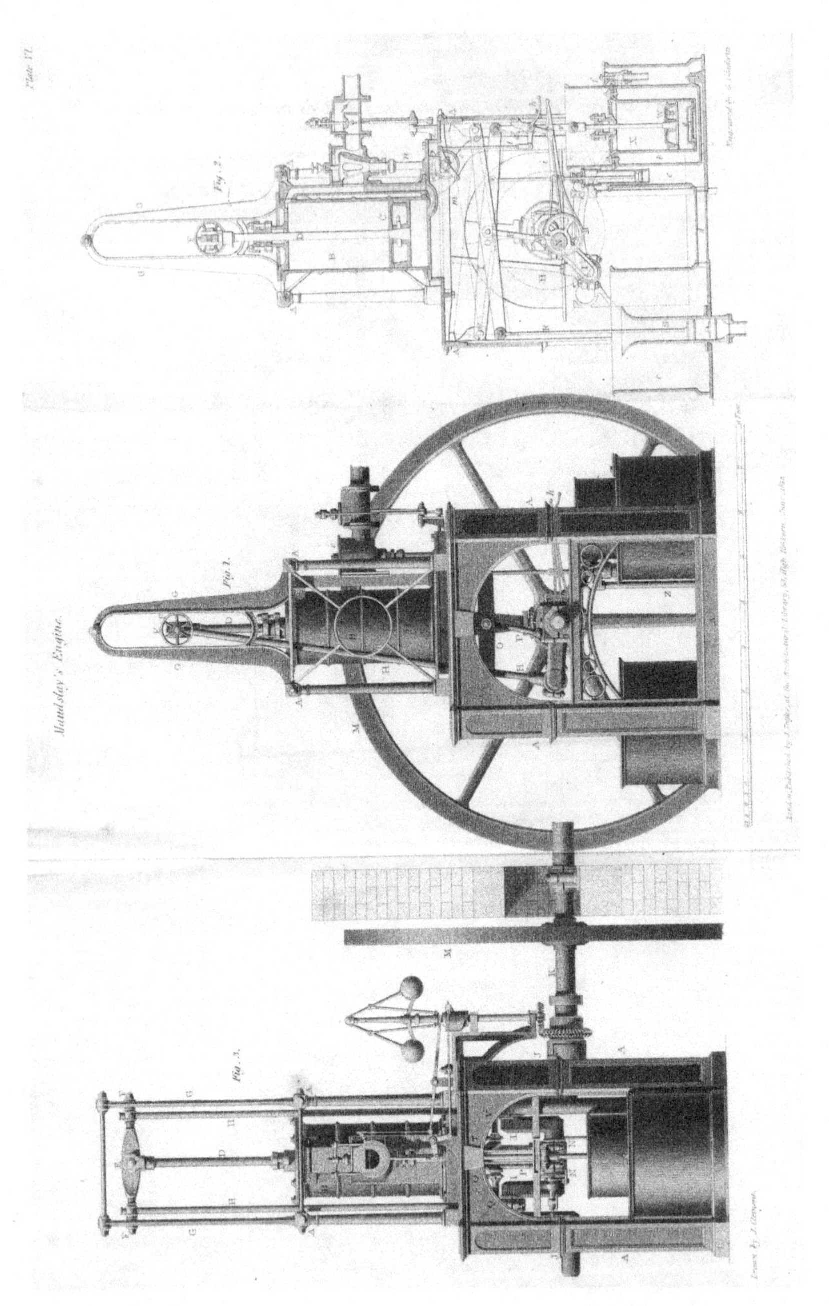

The material originally positioned here is too large for reproduction in this reissue. A PDF can be downloaded from the web address given on page iv of this book, by clicking on 'Resources Available'.

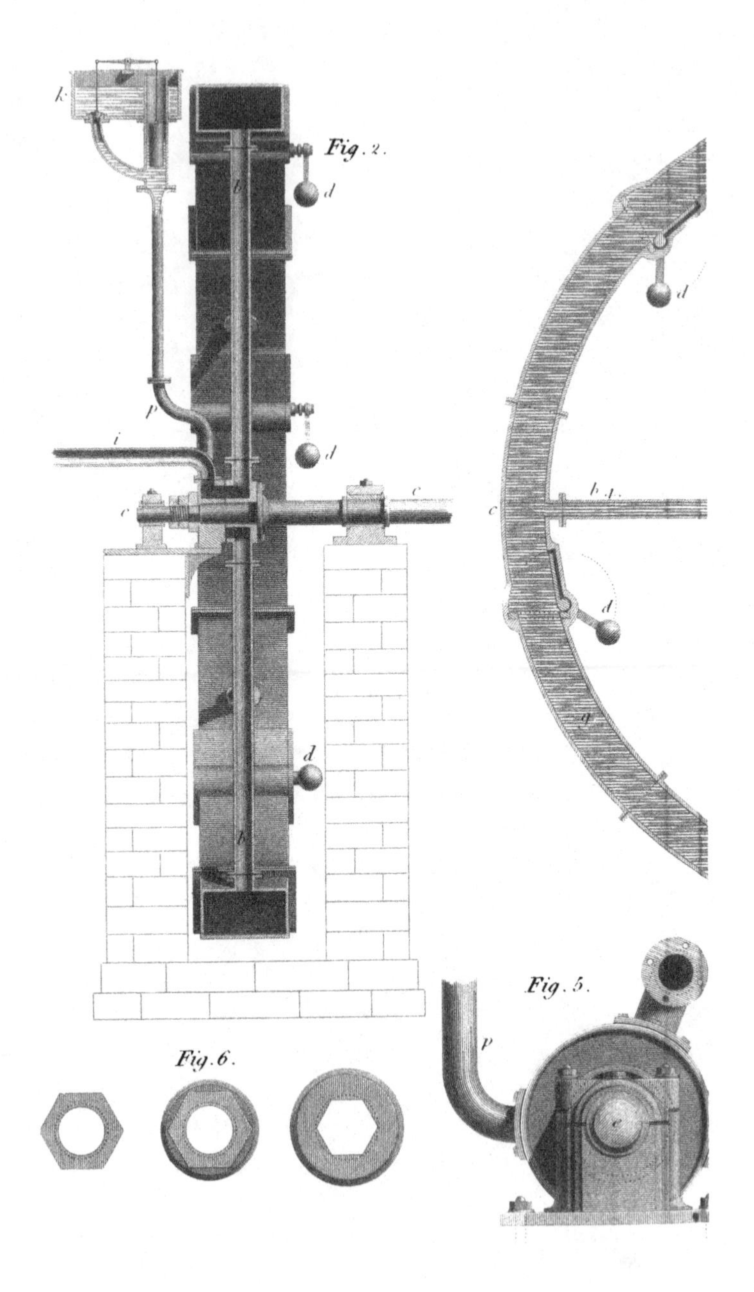

Drawn by J. Clement.

London, Published by J. Taylor, at the Architectural

Rotatory Engine. Plate: VII.

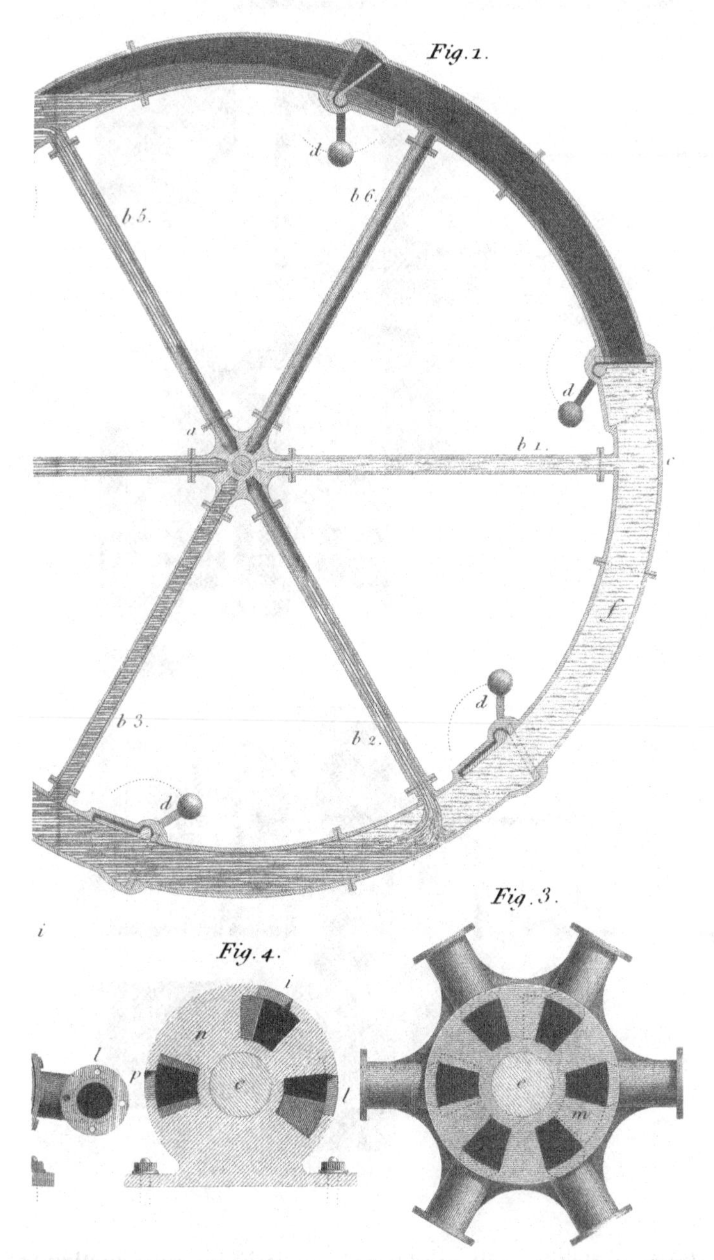

Engraved by G. Gladwin.

Library, 59 High Holborn, June 1822

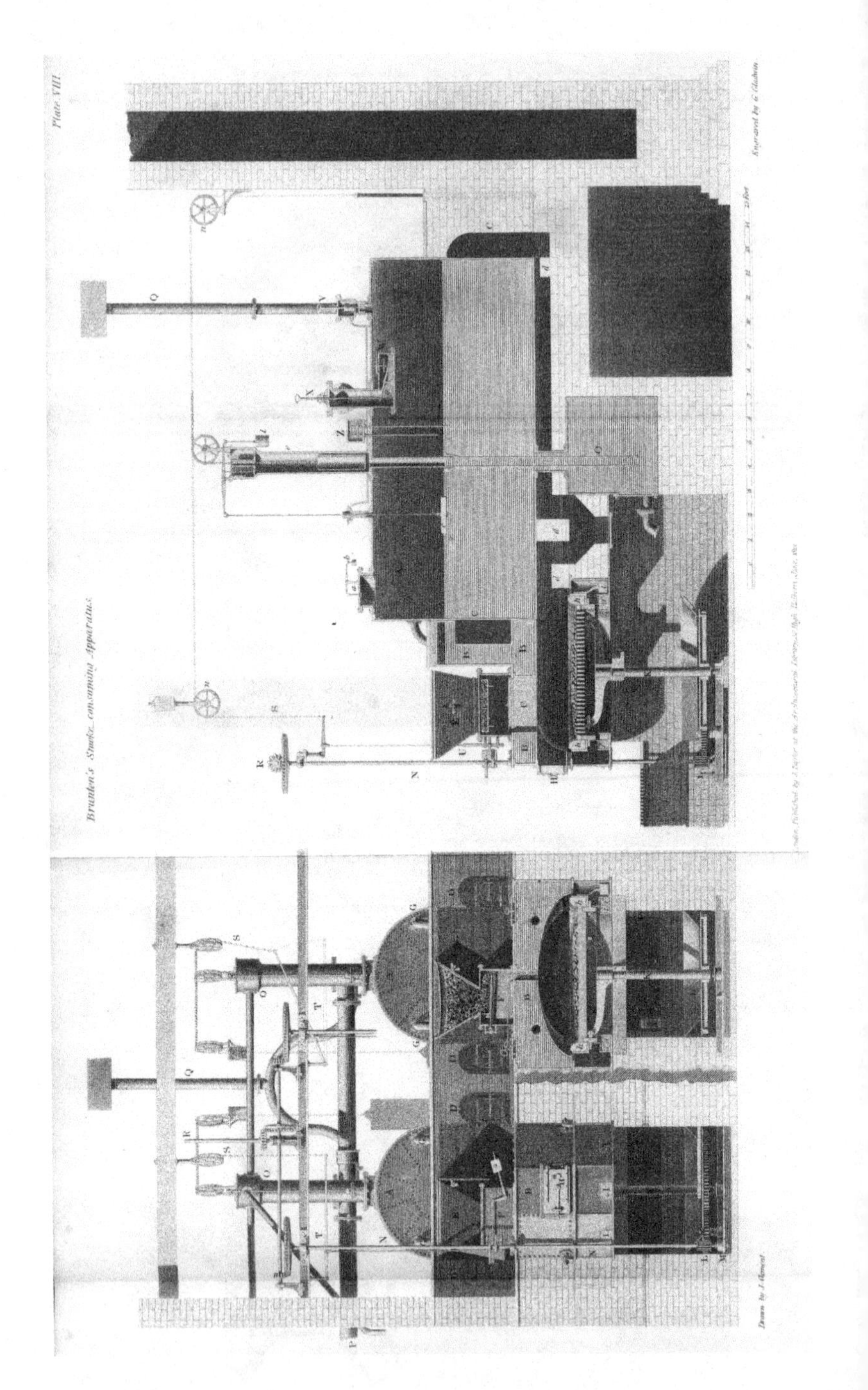

Plate VIII.
Brunton's Smoke consuming Apparatus.
Drawn by J. Clement.
Engraved by G. Gladwin.

The material originally positioned here is too large for reproduction in this reissue. A PDF can be downloaded from the web address given on page iv of this book, by clicking on 'Resources Available'.

Printed in the United States
By Bookmasters